THE FOUR R's

RECOVERY
RECYCLING
RECLAIMING
REGULATION

Richard Jazwin

Business News Publishing Company
Troy • Michigan

Editor: Joanna Turpin
Production Coordinator: Mark Leibold
Cover Art: World Illustration Courtesy Dynamic Graphics © 1990.

Library of Congress Cataloging in Publication Data

Jazwin, Richard.
The Four R's: recovery, recycling, reclaiming, regulation/ Richard Jazwin.
p. cm.
ISBN 0-912524-70-7
1. Fluorocarbons--Environmental aspects--Handbooks, manuals, etc. 2. Air-Pollution--Law and legislation--United States--Handbooks, manuals, etc. 3. Refrigerants--Recycling--Handbooks, manuals, etc. 4. Air conditioning--Environmental aspects--Handbooks, manuals, etc. I. Title.
TD887.F67J38 1992 92-7588
363.73'84--dc20 CIP

Printed in United States
7 6 5 4 3 2

DEDICATION

To the contractors and technicians of tomorrow.

DISCLAIMER

CONTENTS

THE FOUR R'S
RECOVERY
RECLAIMING
RECYCLING
REGULATION

INTRODUCTION

Future service requirements for the refrigeration industry have changed and will continue to change. Recent scientific discoveries have established that certain refrigerants are harmful to the atmosphere - especially to the ozone layer. These discoveries mean contractors and technicians in the hvac/r and automotive fields will have to revise certain accepted field service procedures. And, to implement these revisions, new equipment will have to be purchased and new methods of handling refrigerant instituted.

Scientists now know the ozone layer over the earth is breaking down, and this breakdown will permanently change the hvac/r industry as we know it. One of the major contributors to this breakdown is the release of CFCs into the atmosphere. Since the ozone layer acts as a shield from excess ultraviolet radiation, any decrease in this shield will create problems for the earth and mankind. For years, service technicians have routinely released refrigerant into the atmosphere. These uncontrolled releases have set in motion certain chemical processes, which are causing the progressive deterioration of the ozone layer.

With the scientific evidence showing CFCs are the prime culprits in the ozone breakdown, laws are being passed that affect the hvac/r industry. Did you know that certification will be required for anyone working on refrigeration systems? In addition, reclamation equipment will become a part of everyone's tool kit. Every legislator in the country is trying to do something about the refrigerant problem. These legislators have tried everything from imposing a tax on CFCs to outright banning of automobile air conditioning in certain states.

In the 1970s, the energy crisis caused the government to establish standards for fuel use. The government's action, along with the automotive manufacturers enforced cooperation, resulted in cars that achieved unheard of miles per gallon. The truth is, the government drove the marketplace and that drive was successful.

In the refrigeration area, the government is moving again. This time, the movement is very rapid and the laws that are coming off the government presses may, in fact, either not be sufficient, or worse yet, create more problems than they solve. There are no quick fixes to the ozone problem. Laws of physics cannot be legislated out of existence, and there is no doubt that something must be done. There is a need for control, but let us all hope that the controls imposed do not create impossible standards for compliance. Comfort cooling and refrigeration are not luxuries - they are necessary for mankind to survive in many different environments.

An unfortunate part of the legislative approach to technical problems is that not everyone who authors a bill understands the laws of nature that govern the refrigeration process. Of equal importance, a certain portion of our legislators cannot seem to get past the revenue line when it comes to refrigerants. The current and future taxes that have been imposed on the CFCs will raise millions of dollars, but not one of those dollars is targeted to aid the environmental cause.

Let the reader beware, regardless of the confusion that currently exists, there are laws in existence and these laws will not go away. If anything, the laws are becoming more complex and the confusion on the part of the refrigeration industry more widespread.

This book will help the reader understand the basics of environmental problems and the link to refrigeration. At no other time in the refrigeration industry has there been such a tremendous need for education, training and most importantly, understanding. The reality is that the ozone problem is real. If we do not help to slow the ozone layer deterioration, disaster will result.

CHAPTER ONE

HOW REFRIGERANT AFFECTS THE OZONE LAYER

The ozone layer is breaking down, and this breakdown will affect our world. To service technicians and anyone else who handles air conditioning or refrigeration equipment, this means a drastic change in the way we perform service. New equipment will have to be purchased, certification will be mandatory, new service procedures will be established, and concern for the environment will become uppermost in every technician's mind.

Before we explore what is happening to our world, however, we must first understand a little basic science. Don't panic - it's not that complicated. Most of what we are going to discuss is just a matter of simple addition and subtraction.

MATTER AND MOLECULES

Matter is composed of molecules. A molecule is the smallest possible particle of an element or compound that still retains the properties of that compound. The building blocks of molecules are atoms, and atoms are particles that combine to form a molecule.

A molecule of oxygen contains two oxygen atoms, and it is written O_2. The little number 2 tells us that there are two oxygen atoms forming a gas we know as everyday oxygen. Take away one of the oxygen atoms and we no longer have ordinary oxygen - we have two separate atoms of oxygen.

Ozone, the gas we are concerned with, has 3 oxygen atoms and is written as O_3.

> The three forms of oxygen are:
> Atomic oxygen = O (one atom).
> Ordinary oxygen = O_2 (two atoms).
> Ozone = O_3 (three atoms).

OZONE AND THE OZONE LAYER

About 7 to 25 miles above the earth is a zone called the stratosphere. The stratosphere contains 90% of the earth's ozone in a layer called the ozone layer. The ozone contained in the ozone layer is created by a natural process involving the sun.

The sun emits ultraviolet radiation. Energy from this ultraviolet radiation splits ordinary oxygen molecules (O_2). (Remember, ordinary oxygen contains two oxygen atoms (O_2).) When the oxygen molecule is split, the result is two separate atoms of oxygen (O). The split oxygen atoms combine with other oxygen molecules (O_2) to form unstable oxygen called ozone (O_3). This is how ozone is created.

The ozone layer acts as a shield, protecting the earth from too much exposure to the sun's ultraviolet radiation. The ozone layer literally absorbs some of the sun's ultraviolet radiation, and this absorption creates ozone.

THE CREATION OF OZONE
1. Starts with oxygen (O_2), then splits = Two free oxygen atoms (O).
2. One free oxygen atom (O) bonds with oxygen molecule (O_2) = Ozone (O_3).
3. Result = Ozone with 3 oxygen atoms.

Ozone results because the oxygen molecule absorbs energy from the sun's ultraviolet radiation. The ultraviolet radiation supplies the energy necessary to start the ozone creation process.

Ozone is also present in the atmosphere closest to the earth in a zone called the troposphere. The ozone present in the troposphere is a pollutant. It is caused by ultraviolet radiation acting on smog and air pollutants. Ozone, when present in the troposphere, is not healthy. However, it is not this low level ozone that we are concerned about.

ULTRAVIOLET RADIATION

Ultraviolet radiation is invisible to the human eye, because it has wavelengths shorter than the light we see. Ultraviolet radiation is also known as black light. For centuries, the atmosphere, particularly the ozone layer, has absorbed this ultraviolet light. As we are finding out, though, ultraviolet light or radiation can be harmful to our planet.

As far as we know, the ozone layer has always done an excellent job of absorbing ultraviolet light. In the late 1970s, however, scientists became concerned that the ozone layer was decreasing. The areas in the ozone layer in which there is a decrease of ozone are called *holes*.

As a result of the holes in the ozone layer, more ultraviolet light now reaches the earth. This will cause:

- More cases of skin cancer
- More cases of cataracts
- Immune system weakening
- Increase of ground level ozone, an ingredient of smog
- Crop damage
- Damage to marine organisms
- Increase in global warming
- Rising sea levels

Any of the above individually constitute a problem for the earth. Collectively, at best, these deviations will cause a significant change in our lifestyle. At worst disaster!

REFRIGERANTS AND OZONE

Over the past twenty years, the scientific community has been conducting tests to find out why the ozone layer is decreasing. As research efforts increased, chlorofluorocarbons (CFCs) became the prime suspects for the decrease in the ozone layer.

Chlorofluorocarbons (CFCs) - The Prime Suspects

Anyone in the refrigeration industry deals with chlorofluorocarbons on a daily basis. A *chlorofluorocarbon* contains chlorine, fluorine, carbon and sometimes hydrogen. *Halogenated refrigerants* are refrigerants that contain a halogen. Halogens are any of the five non-metallic chemical elements: fluorine, chlorine, bromine, astatine or iodine.

Fully halogenated CFCs do not contain hydrogen. It is the fully halogenated refrigerants that have the greatest effect on the atmosphere. The term fully halogenated refers to a refrigerant that contains halogens. R-12 is a good example of a fully halogenated CFC. R-12 contains chlorine and fluorine, and it is one of the CFCs responsible for the ozone layer depletion.

The discharge of CFCs to the atmosphere starts a chain reaction, which results in less ozone. In the atmosphere, a chlorine atom breaks off from a CFC molecule. This chlorine atom is now free and invades an ozone molecule. This invasion causes the ozone molecule to break apart. The ozone molecule is destroyed, and we are left with a chlorine atom and an oxygen atom, which form a molecule of chlorine monoxide. Along comes an oxygen atom and breaks apart the chlorine monoxide molecule. We end up with a free chlorine atom, which starts the whole process all over. During the process, a molecule of ozone is lost.

CFCs can live in the atmosphere for over 100 years. One chlorine atom has the potential to destroy over 100,000 ozone molecules. The potential for future ozone destruction is enormous.

HFCs and HCFCs

When a CFC contains hydrogen, it is not fully halogenated. The addition of hydrogen causes the refrigerant to become a hydrochlorofluorocarbon, or HCFC. R-22 is a HCFC. If there is no chlorine, the refrigerant is a hydrofluorocarbon, or HFC.

Current theory states that the non-fully halogenated refrigerants have a lower potential for ozone depletion. They have a shorter life span in the atmosphere, and this creates less potential for ozone depletion. Because of their lower ozone depleting potential, there is less concern about the HCFCs. As research continues, more emphasis is being placed on the non-fully halogenated refrigerants. Ozone depletion is occurring at a faster rate than previously believed, however, so timetables for

HCFC phaseouts and service controls will probably be accelerated.

GREENHOUSE EFFECT
There is much in the news today on another phenomena called the *greenhouse effect*. It is believed the greenhouse effect is causing a rise in temperature on earth. As sunlight reaches the earth, some of the resultant heat normally escapes back into the atmosphere. Proponents of the greenhouse effect believe that there is a slowdown in heat returning to space, which is causing the temperature to rise.

The use of fossil fuels is one cause of the greenhouse effect, and we use many different kinds of fossil fuels. The gases emitted by these burned fuels build up in the atmosphere. The sun rays can still pass through, but the heat that normally would be returned to space is trapped by the gas buildup. Increased usage of fossil fuels results in increased buildup. More heat is then trapped and the earth temperature rises. The scientific community is beginning to suspect that the CFCs may also be a contributor to this effect. However, all the facts are not yet in.

MONTREAL PROTOCOL
In the 1970s, scientists became aware of the ozone layer deterioration. Many scientists at that time began to suspect chlorine as a cause. Subsequent experiments showed that chlorine was, indeed, a leading cause of the holes in the ozone layer.

CFCs have been used for many things other than refrigeration. CFCs are an excellent chemical for aerosol propellant purposes, and they are also used as a cleaning agent. When chlorine drawbacks became accepted, the United States Federal Government began to get involved. The Environmental Protection Agency (EPA) and the Food and Drug Administration (FDA) reacted by banning the use of CFCs in aerosol propellants. This ban was created because aerosols accounted for 50% of CFC usage. Unfortunately, the whole world did not follow suit, and many nations continued to use CFCs as aerosol propellants.

World attention began to focus on the problems caused by CFCs, and it was determined that these problems could not be solved by one nation. In a 1987 meeting in Montreal, Canada produced the *Montreal Protocol*. The Montreal Protocol mandated that CFC production be cut back and that levels be established that acted as baselines for production. The ultimate goal of the Protocol is to phase out production of those CFCs that are harmful. The CFCs that were first covered were R-11, 12, 113, 114, and 115. The Montreal Protocol brought the CFC problem to a global level.

The Protocol was signed by 30 countries. Currently, the United States, the European Economic Community, and fifty other nations have signed the Protocol. The Protocol also establishes a future timetable for meetings to review the impact of science and technology on the environment.

In 1988, the EPA established the provisions of the Montreal Protocol as regulations for the United States. This action was the beginning of the regulatory process for the air conditioning and refrigeration industry.

In 1990, a new law entitled the *Clean Air Act* was passed. This law will impact the entire refrigeration industry. This book will primarily focus on the provisions of the Clean Air Act.

CHAPTER TWO

THE CLEAN AIR ACT

The most significant piece of legislation to affect the air conditioning and refrigeration industry in the United States thus far is the *Clean Air Act* (CAA). The CAA was signed by President George Bush on November 15, 1990. As stated in the previous chapter, a lot of the rules in the CAA resulted from the Montreal Protocol.

The Montreal Protocol attempted to deal with the environmental problems created by refrigerants on the international level. The CAA deals with those same problems, but on the national level. The Montreal Protocol is structured so that periodic meetings must take place in order to reassess the ozone problem. As new facts about the impact of refrigerants are unearthed, the Protocol will be modified. The majority of the Protocol's modifications will also result in the CAA being modified. Language already exists in the CAA stating that the EPA can accelerate phaseout schedules if the EPA deems it necessary. The CAA also states that accelerated phaseout will occur if required by the Montreal Protocol.

The CAA is more specific than the Protocol in addressing the ozone depletion problem. The Clean Air Act gives the EPA the authority to establish environmentally safe procedures with respect to the use and reuse of refrigerants. In addition, the EPA has established standards for certification and service of refrigeration equipment and personnel. These standards have been derived from the information furnished mainly by private sector organizations working in concert with the EPA.

The CAA also establishes a timetable for the phaseout of certain refrigerants and chemicals. The initial approach is to limit refrigerant production on a calendar basis. This means production will be cut back a certain percentage every year until total phaseout of certain chemicals takes place. For example, the CAA calls for the phaseout of CFCs by the year 2000. This phaseout has been accelerated, so that the production of R-12 will now cease at the end of December, 1995. Data obtained from research conducted by various organizations will be the determining factor as to whether other phaseouts will occur sooner.

TITLE VI-STRATOSPHERIC OZONE PROTECTION

In the CAA is a section called *Title VI-Stratospheric Ozone Protection*. Title VI establishes regulations for the production and use of CFCs, Halons, HCFCs and HFCs. Chemicals such as carbon tetrachloride are also covered, however, they will not be treated in this book. Title VI breaks the substances to be regulated into two classes - Class I and Class II.

Class I substances are those substances that cause significant harm to the ozone layer. These substances are broken down into five groups, Table 2-1. In addition, their ozone depletion potential is measured on a scale. Ozone depletion potential (ODP) is just what the name implies. It is the ability to measure, on a scale, the potential of a substance to deplete ozone. Substances having an ozone depletion potential greater or equal to 0.2 are considered Class I, Table 2-1.

Table 2-1. Class I Groups

GROUP I	R-11, 12, 113, 114, 115
GROUP II	HALON 1211, 1301, 2402
GROUP III	Other CFCs having 1 to 3 carbon atoms
GROUP IV	Carbon tetrachloride
GROUP V	Methyl chloroform except 1,1,2 isomers

Note : An isomer is defined as any two or more chemical compounds that contain the same number of atoms of the same element but differ in structure and properties.

Class II substances are those which are known to cause, or are suspected of causing, harm to the ozone layer. Class II includes all isomers of HCFCs that have 1, 2 or 3 carbon atoms. R-22 is a Class II substance.

Ozone depletion potential is the measure of a chemical's ability to destroy ozone molecules. Some sample ODPs are shown in Table 2-2.

Table 2-2. Ozone Depletion Potential

R-11	1.0 Depletion Weight
R-12	1.0 Depletion Weight
R-113	0.8 Depletion Weight
R-114	1.0 Depletion Weight
R-115	0.6 Depletion Weight
HALON 1211	3.0 Depletion Weight
R-22	0.05 Depletion Weight

Production and Phaseout of Certain Refrigerants

The CAA establishes a timetable that sets maximum allowable production of Class I substances and the date for total phaseout. For example, CFCs and Halons are considered Class I substances. As such, they will be phased out on a calendar basis until total phaseout occurs on January 1, 2000. If continuing research determines that ozone depletion is substantially more than originally indicated these timetables will be accelerated.

The CAA established a production baseline for R-11,12, 113, 114 and 115. Effective January 1st of subsequent years, production of these CFCs is to be decreased until phaseout occurs, Table 2-3.

Table 2-3. Original Refrigerant Production Baseline for R-11, 12, 113, 114 and 115

YEAR	ALLOWABLE PRODUCTION
1991	85%
1992	80%
1993	75%
1994	65%

(continued on next page)

Table 2-3. Continued.

1995	50%
1996	40%
1997	15%
1998	15%
1999	15%
2000	0%

Note: A change to the original phaseout schedule is that the production of R-12 will cease on December 31, 1995.

By the year 2000, if not sooner, production of these refrigerants will be non-existent. As an additional incentive to discontinue use of certain ozone-depleting chemicals, taxes will be imposed on those who purchase these refrigerants - namely, wholesalers and subsequent users. The concept of adding taxes to discourage consumption is a good one. However, it is interesting to note that these additional taxes are not going into any type of fund that researches or solves the ozone depletion problem.

Class II substances, HCFCs, are also on a timetable. Effective January 1, 2015, production of Class II substances will be frozen and use limited to refrigeration. It will be possible to use Class II substances in new refrigeration equipment until January 1, 2020. After January 1, 2020, the only permitted service will be on "in place" equipment, and a total production ban becomes effective January 1, 2030. It is possible that all of the production dates shown will change. R-22 production in the U.S. may cease as early as the year 2005.

While all of the above dates appear distant, remember, these dates can change. An equally important consideration is the life of refrigeration equipment. A good household refrigerator can last 20 years. If the unit is purchased in 1995 and uses R-12, a repair requiring R-12 refrigerant in the year 2000 will not be possible. It is hoped that acceptable substitutes will be readily available by the timed phaseout implementation.

REFRIGERANT DISCHARGE

There are provisions in the CAA that will affect refrigeration service. One of these provisions deals specifically with the subject of venting or discharging certain refrigerants into the atmosphere. This provision states:

Effective July 1, 1992, it shall be unlawful for any person, in the course of maintaining, servicing, repairing, or disposing of an appliance or industrial process refrigeration, to knowingly vent or knowingly release or dispose of any Class I or Class II substance used as a refrigerant in such appliance (or industrial process refrigeration) in a manner which permits such substance to enter the environment. De minimus releases associated with good faith attempts to recapture and recycle or safely dispose of any such substance shall not be subject to the prohibition set forth in the preceding sentence.

That is the language used in the House Senate Conference report on the CAA. R-12 is a Class I refrigerant, R-22 is a Class II refrigerant. Do not be fooled by the term appliance. A rooftop refrigeration unit is a stationary appliance. Automotive air conditioning has its own specific requirements. Discharging refrigerants is a thing of the past.

Effective July 1,1992, You cannot legally discharge refrigerant to the atmosphere. Effectively, you must recover, reclaim or recycle. The day of letting refrigerant charge blow off is over. The exception is an accident in the process of trying to capture the charge.

The CAA required the capture and recycling of CFC and HCFC refrigerants by July 1, 1992. Recycling equipment must be a part of everyone's equipment requirements in order to legally service refrigeration equipment.

AUTOMOBILE AIR CONDITIONING

One of the largest users and losers of refrigerant is the automobile air conditioning system. The primary refrigerant used in automobiles is R-12. The CAA has new standards for automotive air conditioning service, which state a technician must:

- Be properly trained
- Be certified
- Use recycling equipment that is certified

If an auto repair shop serviced more than 100 automobiles in the calendar year of 1990, then the CAA said all of this must take place by January 1, 1992. If a repair shop serviced less than 100 autos in 1990, then the compliance date was extended to January 1, 1993.

The above requirements are for CFCs, but mandatory certification for HFC use comes into effect November 15, 1995.

ENVIRONMENTAL PROTECTION AGENCY

The CAA was passed in order to begin solving the environmental problems created by CFCs and other chemicals. The CAA provides for the establishment of many new procedures and standards, and these standards have been issued by the Environmental Protection Agency (EPA). See Appendix A for the actual regulations.

The new standards come in the form of regulations issued by the EPA. The majority of these regulations are based on data supplied by industry groups, committees and technical advisors from both government and private sector organizations. Many of the new regulations already existed in the form of technical papers. For example:

- Standards for automotive air conditioning service were derived from the Society of Automotive Engineers (SAE) papers on the subject.
- Standards for the minimum acceptable purity for recycled refrigerant were derived from the SAE and/or the Air Conditioning and Refrigeration Institute.

It is the function of the EPA to set these standards as regulations. It will be the function of future contractors, repair facilities and technicians to implement this new way of doing business. Over the next five years, a tremendous amount of cooperation will be necessary to ensure that the standards and procedures are not detrimental to personnel or equipment.

ALTERNATIVE REFRIGERANT

The EPA is required to list safe alternatives for the Class I and Class II substances covered in the CAA. This listing must be published within two years of the CAA enactment. In addition, the EPA will require the use of these alternatives in the federal government projects by May 15, 1993.

OTHER MATTERS

CFCs and HFCs are used outside of refrigeration, often for non-essential purposes. An example of a non-essential purpose is a noise horn. Under the CAA, CFCs were banned for non-essential uses on November 15, 1992. HCFCs used in foam production (with the exception of insulation) will likewise be banned by January 1, 1994.

According to CAA requirements, substances that pose a threat had to have warning labels on their transport containers by May 15, 1993. If a product is made from one of the regulated chemicals, it must also carry a warning label by January 1, 1994.

In many cases when a federal law is passed, individual states deal with the problem that the federal law addresses. This usually means the state is trying to create a law that better suits the state's needs. The state law then preempts (replaces) the federal law in that individual state. This procedure is called preemption.

There are many issues that the CAA covers. Time deadlines have been established for the implementation of phaseout, automotive technician certification, and recycling equipment standards. All of these new standards must be met.

The EPA published a summary of the final regulations in June, 1993. Prior to publication of the official regulation, a *Notice of Proposed Rulemaking* for public comment was issued. The reader should understand that regulation is about to become a way of life. Time deadlines and standards will change as the environmental impact of chemicals becomes better understood. It is fairly safe to say that timetables calling for events to happen by the year 2000 will be accelerated. The truth is, as research continues, the impact of chemicals such as CFCs is even greater than previously suspected.

So far we have briefly covered recovering, reclaiming, and recycling. Don't worry, there is still much more to come on these topics. However, there is a fourth R - *regulation*. This is what will change all of our lives.

CHAPTER THREE

AN OVERVIEW OF CERTIFICATION

Industry experts say that 25% of the R-12 purchased in the United States is used in automobile air conditioning. In addition, the hoses and fittings used in automobile air conditioning allow leakage to occur. R-12 leakage is one cause of ozone layer deterioration. In view of these facts, the automotive industry is one of the first industries for regulation under the CAA.

The CAA grants the EPA the responsibility of creating regulations to improve the ozone situation. Therefore, the EPA has the authority to implement regulations that will establish standards for the use of certain refrigerants.

One of the first targets for regulation under the CAA is automobile air conditioning. The EPA must establish requirements for training and certification of automobile air conditioning technicians. The time frame for technician training, equipment approval and mandatory service procedures is cited in the CAA.

The CAA, via the EPA, has established certification requirements for hvac/r personnel. These requirements are spelled out in EPA regulations. Effective November 14, 1994, all hvac/r service personnel must be certified. The rule also states that it is unlawful to vent or discharge refrigerants into the atmosphere, and this definitely applies to hvac/r personnel.

AUTHORITY

Currently there is a lot of confusion in industry as to what is required under the CAA. The following paragraphs are taken directly from the Clean Air Act. The CAA in Section 7671 h paragraph (d) Certification states the following:

1. Effective 2 years after the enactment of the Clean Air Act Amendments of 1990, each person performing service on motor vehicle air conditioners for consideration shall certify to the Administrator either-

(A) that such person has acquired, and is properly using, approved refrigerant recycling equipment in service on motor vehicle air conditioners involving refrigerant and that each individual authorized by such person to perform such **service is properly trained and certified;** or

(B) that such person is performing service at an entity which serviced fewer than 100 motor vehicle air conditioners in 1991.

2. Effective January 1, 1993 each person who is certified under Paragraph (1) (B) shall submit a certification under paragraph (1) (A).

In Section 7671 (h), the following language defines the terms properly trained and certified:

The term properly trained and certified means that training and certification in the proper use of approved refrigerant recycling equipment for motor vehicle air conditioners in conformity with standards established by the Administrator and applicable to the performance of service on motor vehicle air conditioners. Such standards shall, at a minimum, be at least as stringent as specified, as of November 15, 1990, in SAE standard J-1989 under the certification program of the National Institute for Automotive Service Excellence (ASE) or under a similar program such as the training and certification program of the Mobile Air Conditioning Society (MACS).

What It All Means
If you service automobile air conditioners, you will have to be certified. In addition, recycling equipment meeting minimum standards will have to be purchased. Finally, certified technicians will have to understand the basics of recycling equipment. The regulation states:

All persons performing service on motor vehicle air conditioners for consideration, notwithstanding the size of the establishment, must certify to the Administrator on or before January 1, 1993 that they have purchased approved equipment and that authorized service personnel have been properly trained and certified.

How Certification Will Work
There are several programs in existence that certify automobile technicians, and more than likely, there will be more. From the EPA's standpoint, a certified automobile technician must:

- Be aware that venting refrigerant is illegal.
- Understand why all the regulations are being created (this is based on the technician understanding what is happening to the environment).
- Have a working knowledge of SAE standards J-1989, J1990 and J1991.
- Perform service in a safe manner, without injuring personnel or damaging equipment. Areas that must be understood include venting, handling, transporting, and disposing of refrigerant.

The CAA states, "Such standards shall, at a minimum, be at least as stringent as specified, as of November 15, 1990, in SAE standard J-1989 under the certification program of the National Institute for Automotive Service Excellence (ASE) or under a similar program such as the training and certification program of the Mobile Air Conditioning Society (MACS)."

The main problem in any certification program is its scope. The EPA's primary goal is to make sure technicians understand the environmental impact of refrigerant. The technician must also understand and adhere to the newly established rules that exist for refrigerant service. In addition, though, many groups feel that approved certification programs should also include testing technical competence of the technician. There exists a division among these groups as to how much training should be required for technician certification. Everyone understands that the law is already in place, but the law is vague when it comes to the quantity of training required.

In the hvac/r field, there is also a very strong movement to establish minimum certification standards for hvac/r service technicians. The EPA was supposed to establish the ground rules for certification and approve a program by January 1, 1992. However, as with the automotive certification, this date was pushed back. Please see Appendix A for the latest information on technician certification.

In the automotive industry, we do know that ASE and MACS are named in the CAA as potential certifying bodies. As time progresses, hvac/r technician certification programs will be approved by the EPA. In order to issue certification for hvac/r technicians, an organization will have to meet rigid standards that have been determined by the EPA.

MACS

MACS stands for the Mobile Air Conditioning Society. MACS was directly involved in formulating standards and regulations regarding refrigerant recycling for the automotive industry. Currently, MACS offers a training program and test for technician certification. This program involves basic training in the environmental, service and safety requirements that should be required for technician certification.

How the Program Works

In the MACS program, a technician becomes certified through a home study program, which includes a booklet and a test. The booklet covers all of the new requirements and describes the ozone problem, identifying the factors that help create the ozone breakdown. The technician studies this booklet, then takes the test - a minimum score of 80% is necessary to pass. It is an open book test, so the technician may refer to the study booklet to determine the proper answer. The completed test is mailed to an independent testing service and graded. Upon satisfactory completion of the test, a certificate and pocket identification card are issued to the technician.

In addition, there is also an on-site proctored test available. The on-site test format offered by MACS includes a training session that occurs prior to the closed book test. The passing score on the closed book test is lower than the test using the mail-in format. The MACS training and testing procedures have been officially approved by the EPA. Technicians who have taken the MACS test prior to EPA formal approval will probably be "grandfathered" into certification.

ASE

ASE stands for the National Institute for Automotive Service Excellence. For many years, ASE has tested automotive technicians for competence in performing service procedures.

Currently, ASE offers certification in automotive air conditioning. This certification is primarily concerned with the technical competence of the technician. Work experience is necessary before a technician can take the ASE certification test. ASE feels that a technician should be technically competent in his or her craft before they can be certified. The completion and certification by ASE in an area such as air conditioning demonstrates to the public that the technician has reached a minimum standard of proficiency in service.

As stated earlier, one of the main concerns in automotive air conditioning is what the minimum standards for certification should be. ASE believes that one requirement for certification should include technical expertise, and the ASE is mentioned in the CAA as having a model certification program. There are many automotive technicians working in the field who already hold an ASE certification in air conditioning. ASE now has a certification program in place that will cover certification under the CAA. ASE's program is also available through the mail and is an open-book test. The ASE program is approved by the EPA. The EPA will set the standards, and the professional associations will then have to tailor programs to meet these standards.

HVAC/R CERTIFICATION
Regulations requiring recovery and recycling of refrigerant became effective July 1, 1992. There is a regulation that requires certification for personnel servicing automotive air conditioners already in place. The last major area to regulate is the hvac/r industry.

Appendix A contains a summary of EPA regulations dated June, 1993. This summary provides the rules and regulations that now apply to the hvac/r industry.

Technician certification is a reality. All technicians working in the hvac/r industry will have to be certified by November 14, 1994.

In order to become certified, a technician will have to take a closed book, proctored test. The test will be offered by organizations that have been approved by the EPA. There will be no training requirement attached to the certification test.

Unlike the automotive program, certification for the hvac/r industry will be offered in levels corresponding to the type of service the technician performs. These levels are listed as *types*.

There are four types of certification:

- Type I - for servicing small appliances.
- Type II - for servicing high or very high pressure appliances.
- Type III - for servicing or disposing of low-pressure appliances.
- Type IV (Universal) - for servicing all types of equipment.

The regulations also cover areas such as:

- Service practice requirements
- Levels of evacuation
- Equipment certification
- Equipment grandfathering
- Refrigerant leaks
- Refrigerant sale restrictions
- Certification by owners of equipment
- Reclaimer certification
- MVAC-like appliances
- Safe disposal requirements
- Recordkeeping requirements
- Hazardous waste disposal
- Enforcement
- Compliance dates

Many technicians have already taken examinations offered by different associations. It was the intent of these organizations to offer the examinations in the hope that the technicians who passed these exams would be grandfathered. Check with the association that offered the exam to see if you are, in fact, grandfathered, or whether or not you will have to take a new exam.

CHAPTER FOUR

RECOVERY

Recovering, recycling, and reclaiming will forever change the way service is performed. It is important that the reader understand the difference between these three terms. Recovery will be explained in this chapter, recycling and reclaiming in subsequent chapters.

RECOVERY DEFINED

"To remove refrigerant in any condition from a system and store it in an external container without necessarily testing or processing it in any way" is how ASHRAE Guideline 3-1990 defines recovery. The recovery process is concerned with removing refrigerant from the system - it is not concerned with the quality or purity of the removed refrigerant.

PAST RECOVERY PROCEDURES

To some degree, recovery has always been a part of the refrigeration industry. For the most part, however, service technicians have not bothered to recover refrigerant. The technician in auto or hvac/r service merely let the charge go into the atmosphere (called "blowing the charge"). These days are gone. The law requires that refrigerant be recovered and contained in an external container. Don't worry - recovering the refrigerant from the system can be a simple procedure.

RECOVERY PROCEDURES

Recovery is a very big part of refrigeration service. As of July 1, 1992, CFC recycling (recycling is defined in subsequent chapters) is required and the discharging of HCFCs into the atmosphere is no longer allowed. This means a technician performing service requiring refrigerant removal is no longer be able to discharge the refrigerant into the atmosphere. Therefore, the refrigerant charge must be recovered and contained.

Recovery will become even more important once the cutbacks in CFC production begin, because these cutbacks will cause refrigerant shortages. For example, R-12 production is being phased out on a timetable (as illustrated in Figure 2-3). There are many units that use R-12 as a refrigerant. Eventually, the quantity of R-12 that can be purchased will be **less** than the market is able to supply, causing a shortage. This shortage of refrigerant will be offset to some degree by reusing and reprocessing the refrigerants taken from existing systems.

Refrigerants in existing systems can be considered as warehouses for future usage. One of the largest makers of refrigerants projects that in the year 2000, about 30% of market demand will be served through conservation efforts, including recycling, recovery and reclamation.

RECOVERING DIFFERENT REFRIGERANTS

Tomorrow's technician must be aware that recovery varies according to the type of refrigerant being used. Different refrigerants created different methods of recovery.

R-11

One of the easiest refrigerants to recover is R-11. R-11 recovery consists of pressurizing the system receiver and pouring the refrigerant into drums. This method works, because R-11 has a boiling point of approximately 76 °F at atmospheric pressure. When recovering R-11, the service technician is handling a bucket of liquid refrigerant. As long as R-11 is below its boiling point, it is in a liquid state.

R-12, R-22 and R-502

R-12, R-22 and R-502 have to be handled differently. Their boiling point at atmospheric pressure is much lower than that of R-11. Many technicians merely evacuate a storage cylinder and let the refrigerant flow from the system's receiver.

Anyone familiar with the refrigeration process knows that there are laws that apply to refrigerant flow - namely, higher pressure overcomes lower pressure. The presence of a vacuum in a storage container creates an ideal situation for moving refrigerant into the container from a pressurized system. The vacuum allows the refrigerant to flow until the pressure in the container is equal to the pressure in the system. Once pressures are equal, the refrigerant will no longer flow.

A recovery method used in the past involved packing the recovery cylinder in ice. This kept the recovered refrigerant at a lower temperature and pressure. The lower temperature and pressure allowed better refrigerant flow.

There were other recovery methods used, but in most cases, these methods were ineffective, and refrigerant was usually lost to the atmosphere. This refrigerant loss is no longer acceptable. It is the presence of recovery equipment that is giving the industry the ability to develop a uniform refrigerant recovery process without fear of refrigerant loss.

RECOVERY EQUIPMENT

Early refrigerant recovery was used as a matter of pure economics - recovering and reusing refrigerant provided cost savings to the end user. Transit systems throughout the United States routinely used machines to capture the charge in bus air conditioning systems. This recovery reduced the amount of refrigerant the bus companies would have to purchase. When the tax on CFCs went into effect, the need to use recovery machines increased even more. Many people realized that using recovery machines could offset the cost of escalating refrigerant prices.

Recovery/recycling machine sales are increasing rapidly. Sales figures show that in early 1991 over 70,000 recovery units had been sold. In 1991, projections showed that over 130,000 additional units would be put into service, and long-term projections show a market for at least 200,000 additional units. Over the next few years, 400,000 recovery systems will be working to help alleviate the ozone problem.

Routine recovery has not yet achieved industry-wide popularity. Many technicians who work on smaller units feel that it is far easier to blow the charge than recover it. Reality for today's service technician is that blowing the charge is no longer a legally or economically sound business practice.

RECOVERY CONCERNS

The increased use of recovery equipment creates some valid concerns. If recovery is performed incorrectly, it can be dangerous. Some areas of concern are:

- Overfilling external cylinders
- Recovering to improper cylinders
- Refrigerant mixing
- Refrigerant contamination
- Oil return
- Storage and handling
- Equipment requirements and cost

Overfilling External Cylinders

Improper use of recovery equipment can create situations that will cause a cylinder to explode. When transferring refrigerant to an external container, keep the following points in mind:

- Never fill the container with more than 80% of its volume with liquid refrigerant. The remaining 20% allows for expansion.
- "To prevent on-site overfilling when transferring to external containers, the safe filling level must be controlled by weight and must not exceed 60% of container's gross weight rating," SAE J1989.

A cylinder that is full of liquid is not subject to the normal pressure-temperature relationship. When overfilled, the cylinder is subject to hydrostatic (liquid pressure) laws. Circumstances can cause the liquid-filled cylinder to become a bomb and explode. **WARNING: FILLING A CYLINDER OVER 80% OF ITS VOLUME WITH LIQUID WILL CAUSE THE CYLINDER TO BURST, UNDER THE RIGHT CONDITIONS. SOMEONE CAN BE KILLED FROM THE RESULTANT SHRAPNEL. THERE IS NO EXCEPTION TO THIS WARNING. BY FOLLOWING THE RULE, THE LIFE YOU SAVE WILL BE YOUR OWN.**

Recovery to Improper Cylinders
There are laws stating which types of cylinders can be used for recovery (approved cylinders will be covered later). It is important to understand that it is illegal and dangerous to use a disposable refrigerant can for recovery, even for a short period of time. Disposable cans were created to be thrown away. Currently, manufacturers are engineering disposable cans that cannot be refilled under any circumstances. These cans will eventually find their way to the market. In the meantime, **do not use a disposable can for anything** (this includes using a disposable as a compressed air holding tank). Disposable cans do not have the required wall strength or rating to be used for storage. Disposables also have been known to rust, from the inside to the outside, without any visible rust signs on the outside.

Refrigerant Mixing
Some recovery machines are designed to only handle R-12, and these machines should not be used on R-22 systems, or any other systems. The machine is just not designed to handle another refrigerant. If a machine is designed to handle multiple refrigerants, it must be cleaned before attempting service on another type of refrigerant. R-12 residue remaining in the recovery machine could end up mixed with R-22. The resultant pressure characteristics would make tuning the system impossible.

Refrigerants cannot be mixed. Mixing of R-12 and R-22 in a system will create problems. *Note: There are certain times when limited mixing takes place. This is not a normal procedure and is used only to solve a specific problem such as low ambient operation. Azeotropes (mixed refrigerants) are not considered mixed, but rather chemically changed so they adhere to a preset standard.*

The alternative refrigerants currently arriving in the marketplace are a potential source of concern. If mixed with another refrigerant in an existing system, some of the alternative refrigerants possess characteristics that will harm the system. If an alternative refrigerant is present in the recovery unit and mixed with another refrigerant, damage may result.

Recovery equipment must be used properly to ensure that any remaining refrigerant is removed from the unit prior to attempting recovery of another type of refrigerant.

Refrigerant Contamination
The purity of refrigerants returned to the system must meet certain standards or the system will suffer. Currently, there are purity standards that exist for both the automotive and hvac/r industries. Contaminants such as air, water and foreign particles must be removed if the refrigerant is to be reintroduced to the system. When recovering refrigerant that is to be reintroduced to the system, certain procedures must be followed to ensure the purity of the refrigerant.

Impure refrigerant reintroduced to a system can cause a number of problems. One such problem is loss of warranty on a new compressor installed in a system where contaminated refrigerant is present. Another problem occurs when air is allowed to remain in the system, making correct charging impossible. At the very least, a system with air present will result in system inefficiencies. Water or moisture will cause havoc and eventual system failure.

Oil Return
During the recovery process, oil can be removed from the system. The quantity of oil removed depends on the system, the recovery unit and the skill of the operator. Care must be taken to ensure there is enough oil remaining in the system for proper operation to take place.

Alternative refrigerant use creates oil concerns. Some of the alternatives use a different kind of oil than that found in existing systems. Improper system functioning results, and in some cases, the introduction of the original system oil results in harm to the system.

Storage and Safe Handling
There are laws governing refrigerant storage. Storage for our purposes is defined as *refrigerant containment whenever the refrigerant is outside of the operating system.*

Only Department of Transportation (DOT) approved containers should be used. These containers are stamped with DOT numbers that demonstrate the safety factors in cylinder construction. Failure to follow safe handling and storage procedures will place the technician in violation of the law. Improper storage and handling will create a life-threatening situation for the technician and others.

EQUIPMENT REQUIREMENTS AND COSTS

Certified recovery/recycle equipment is now a requirement to operate in today's service. For the technician who performs service on refrigerant systems containing different refrigerants, different recovery machines may be necessary. At the very least, multiple approved storage cylinders will become necessary. The service truck of the future will carry approved recovery tanks for all different refrigerants. This approach will avoid unnecessary trips to the shop to remove recovered refrigerant before the next recovery can be made.

In the automotive field, the new law provides for shutoff valves on the end manifold gauges. The new alternative refrigerants will also require leak detectors to locate leaks. The automotive manufacturers are creating new valves for system entry, therefore new service tools will be required.

The era of regulation is just beginning. Leakage to the atmosphere, illegal discharge and safe service procedures are uppermost in every legislator's mind. New equipment will be developed to meet the needs of the new procedures. In the end, service will become more expensive, and this cost increase will be passed on to the customer. It represents the price that must be paid to help clean up the environment.

One major change in recovery is that used refrigerant cannot be sold to another entity. You can recover and recycle refrigerant, but you must return it to the same system or a system owned by the same entity. **You cannot resell used refrigerant.**

Under the new regulations, recovery/recycling equipment must be capable of pulling a measured vacuum. The EPA summary chart in Appendix A shows the value that must be achieved.

The next chapter will explore the mechanics of recovery for both the automotive and hvac/r industry. The technician of tomorrow will have to be better trained and more environmentally aware. This increase in awareness is what we as an industry must contribute as our share of the environmental cleanup.

CHAPTER FIVE

THE MECHANICS OF RECOVERY

Recovery consists of removing the refrigerant from a system and storing the refrigerant in an external container. The external container may be part of a dedicated recovery machine or a freestanding storage cylinder.

During recovery, the technician is not concerned with the purity of the refrigerant. Normally, all the technician is concerned about is removing the refrigerant from the unit so that service can be performed. When the condition of the refrigerant becomes a concern (for example, if a burnout occurs), then the technician must ensure that the refrigerant is recycled or reclaimed before it is returned to the system. Recycling and reclaiming refrigerant will be covered later, but for now, let's look at what is involved in the recovery process.

RECOVERY PROCESS

Refrigerant can be recovered in either vapor or liquid form. In vapor recovery, refrigerant in the system's evaporator is drawn off by means of a compressor in the recovery machine. The recovered vapor will be condensed in the recovery machine and stored in a storage container.

Liquid recovery can be as simple as drawing a vacuum on a storage cylinder, opening some valves and allowing the system's refrigerant to flow into the storage cylinder. However, with today's refrigerants, there is a good chance that not all of the refrigerant will be removed from the system using this method. Any refrigerant remaining in the system cannot legally be vented, so all remaining refrigerant must be removed some way.

The best method to ensure total refrigerant removal is to use a recovery machine. Using a recovery machine to recover the system's refrigerant gives the technician control of the recovery process and ensures total recovery. Many recovery machines have the capability of recycling as well as recovering refrigerant. If the recovery machine has the capability of recycling, then the machine may also be able to:

- Purge non-condensables
- Remove moisture
- Remove acid
- Remove particulates
- Recover oil from the system

Technology in the recovery area is proceeding quickly. Patents are being issued, which will allow the machines to accomplish safer and faster recovery. Unfortunately, a good portion of the new technology may not be in the market by the date mandated for recovery.

VAPOR VERSUS LIQUID RECOVERY

Vapor recovery is slower than liquid recovery. Liquid recovery should be the technician's first choice, however, there are many factors to consider. Some of these factors are: equipment availability, system operation, and quantity and boiling point of the refrigerant. Many times it is possible to recover liquid first and then finish the recovery process in vapor form. By employing liquid recovery first, recovery will take a shorter amount of time. After the maximum quantity of liquid is recovered, then vapor recovery will finish the job.

MECHANICS OF VAPOR RECOVERY

The procedures given in this section are general in nature and are intended to give the reader an understanding of what the recovery process entails. Recovery machines that possess additional technology that eliminates or shortens a given procedure are even better.

Figure 5-1 is a simple flow chart that outlines the basic recovery operation. Remember, we are dealing with recovery only in this chapter - the technician is not concerned with the purity of the recovered refrigerant. Only after recovery is complete and the system repairs have been made will the technician decide whether or not to return the original refrigerant to the system.

The flow chart and following descriptions involve a straight vapor recovery and are general in nature. There are no directions to open this valve and close that valve. Each recovery machine has its own piping structure and **the user must follow the manufacturer's operating instructions**. Failure to follow the manufacturer's instructions can result in liquid slugging of the recovery unit compressor. As with an air conditioning compressor, a recovery unit compressor is unable to pump liquid refrigerant.

The following steps are involved in vapor recovery:

1. Connect the recovery machine to the refrigeration system according to manufacturer's installation instructions.

2. When attaching the recovery machine to the low side of the system, place a manifold gauge set in the line. In addition, a filter drier, referred to as a pre-filter, can also be installed in the low side. In the event you intend to reuse the refrigerant, the pre-filter will assist in cleaning and drying the refrigerant. *Note: Do not be concerned if the recovery machine is equipped with its own gauges. Installing manifold gauges will not affect the*

recovery unit operation. Placement of manifold gauges will allow the user to isolate the system being serviced from the recovery unit. Isolation allows additional checks to be made on recovery progress. In addition, the manifold gauges can create a platform for later charging of the system.

3. Vapor is drawn out of the system's evaporator through the manifold gauge set and into the recovery machine. The system metering device feeds refrigerant into the recovery unit as the recovery unit compressor draws refrigerant from the system evaporator.

The system evaporator gets colder as the refrigerant vapor is drawn off. This coldness is normal since the recovery unit compressor pulls down the temperature and pressure of the system evaporator. The temperature-pressure relationship still applies; that is, the lower the pressure the lower the temperature.

Figure 5-1. Simple Recovery Cycle

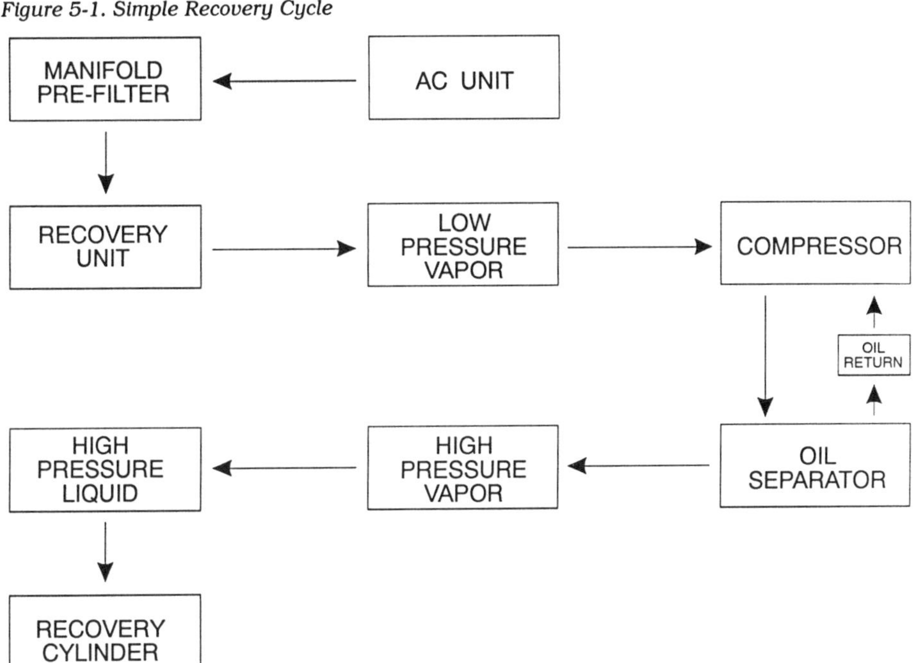

4. The refrigerant entering the recovery unit is piped to the compressor. The recovery unit compressor changes the low-pressure gas to high-pressure gas. The refrigerant then leaves the compressor as high-pressure gas.

5. The high-pressure gas enters the oil separator, where oil is separated, and returned to the compressor.

6. The high-pressure gas condenses to high-pressure liquid and enters the recovery cylinder. Remember, the recovery cylinder should never have more than 80% of its volume filled with refrigerant. There are automatic shutdown systems that will stop recovery at the 80% requirement. These shutdown systems represent a life saving investment for technicians. *Note: Whether the recovery is vapor or liquid, an approved storage cylinder must be used. Current approved cylinders are labeled DOT 4BW, DOT 4BA or UL.*

By now you should understand that the compressor in the recovery machine draws the refrigerant out of the system requiring recovery. If you are recovering vapor, once the recovery process starts, suction pressure will begin to drop. When the gauge reads zero or lower, recovery is complete. Some recovery machines have automatic shutoffs that are set to shut down at 10 to 15 inches hg, indicating recovery completion.

Capacity
Most recovery machines specify that so many pounds of refrigerant per minute can be recovered. These time parameters are not totally accurate and depend on many external and internal factors. Temperature drop is one of these factors. Because of the nature of refrigerant, as the temperature drops, recovery is slower. In vapor recovery, the user is creating a temperature drop.

Recovery in this chapter pertains to small machines designed to recover small amounts of refrigerants. These small machines recover 1/2 to 1 lb. per minute and operate with a compressor of about 1/6 hp. The majority of these machines are designed to handle only R-12.

High-Pressure Warning
Generally, an excessively high pressure indicates the presence of non-condensables in the system. If, after 12 hours, the storage cylinder shows a pressure higher than the corresponding temperature relationship, arrangements must be made to remove these non-condensables. Reusing refrigerants containing non-condensables will result in pressures that will preclude normal system operation. A later chapter

will outline how to check for the presence of air and non-condensables in recovered refrigerant.

Recovery Completion Check

At the end of the recovery process, you must check to make sure all refrigerant is removed from the system. To do this, isolate the system and let the system stand for 5 to 10 minutes. If the system pressure rises to 10 psi, then pockets of cold refrigerant are still in the system. The recovery process must then be performed again. **Any refrigerant that remains in the system must be removed.**

LIQUID RECOVERY

The fastest method of recovering refrigerant is to use liquid recovery. There are no hard and fast rules for when liquid recovery should be used. Many believe that a system charge in the 2-1/2 to 5-lb range can be recovered as vapor. If the system charge is over 5 lbs, however, then liquid recovery is necessary.

As outlined earlier, the simplest method for recovering liquid is to pull a vacuum on the recovery cylinder and let liquid flow from the high side of the system into the recovery cylinder. The pressure difference between the system and the recovery cylinder will allow a substantial amount of liquid to flow in a short period of time.

Some machines have liquid recovery capabilities that allow the user to initially bypass the recovery machine condenser. In this case, liquid flows straight through the recovery machine, bypassing the condenser. When pressure equalizes in the recovery cylinder and the system, the recovery machine, by valve movement, creates a pressure drop in the recovery cylinder. This pressure drop allows additional liquid to flow. When liquid recovery becomes too slow, some valves are reset and recovery is finalized in the vapor mode.

While recovery sounds complicated, the truth is that it is not. It is merely a matter of the operator understanding how to use the recovery machine as the manufacturer designed it to be used.

HVAC/R VERSUS AUTOMOTIVE REQUIREMENTS FOR RECOVERY

The hvac/r and automotive industries each have separate recovery requirements. These requirements are based on the differences found between these two industries.

In *hvac/r recovery*, the following variables influence the choice of recovery machine and method of recovery:

- Hvac/r encompasses a very broad area
- Equipment is larger and systems contain more refrigerant
- Many different refrigerants are used
- Accessibility is necessary (the weight of the recovery machine can preclude the technician from reaching the system with the recovery machine)
- Electrical failure in the system can result in severely contaminated refrigerant
- Systems containing a large charge will take substantially more time to recover than a small automotive system

As compared to the hvac/r industry, the following conditions apply to *automotive recovery*:

- Automotive recovery involves smaller systems
- Amount of refrigerant in the system is less
- Contaminated refrigerant is much less of a problem
- Usually only R-12 refrigerant is used
- Recovery is much faster for smaller automotive systems

The CAA states that servicing of auto air conditioning shall employ standards at least as stringent as SAE Standard J1989. Observance of J1989 will involve changes in the way the automotive air conditioning is serviced and refrigerant recovered.

CHAPTER SIX

RECYCLING

Recycling consists of reducing contaminant levels in recovered refrigerant by using a recycling machine. A recycling machine accomplishes contaminant reduction by oil separation and single or multiple passes through components such as filter core driers. These filter driers reduce moisture, acidity, and particulates in the recovered refrigerant.

Recycling is performed in the field with a recovery/recycling machine. This machine returns the refrigerant to a usable state, so the refrigerant can be returned to the same system.

RECYCLING AND RECLAIMING

Recycling and *reclaiming* are two methods of reducing contamination in refrigerant removed from a system. The basic difference between the two processes lies in the final purity of the refrigerant.

Recycling occurs in the field and removes most contaminants. The recycled refrigerant is then either put back into the system or sent to a reclaim center for further processing. Recycling does **not** return the refrigerant to its original purity. Most small recycling equipment is designed to return the refrigerant to standards set by SAE J1991. SAE J1991 details standards that were established for R-12 refrigerant in automotive use, Table 6-1.

Refrigerants used in hvac/r applications may require additional recycling or reclamation in order to reach a level pure enough for reuse. In the hvac/r industry, the standard for purity is ARI 700-88. In order to reach this standard, reclaiming is necessary. In the recycling process, the technician conducts field tests to determine if the refrigerant is reusable. If the contamination level of the recovered refrigerant is too great, then reclaiming is indicated.

Table 6-1. SAE Standard J1991 Purity in Mobile Air Conditioning R-12.

Contaminant	Allowable Limit by Weight
Moisture	15 ppm
Oil	4000 ppm
Noncondensable (Air)	330 ppm

Reclaiming purifies the refrigerant. Reclaiming involves a distillation process and is performed at a chemical manufacturer's plant or a processor's reclaim plant.

Reclaiming returns the refrigerant to industry standards as called for in the Air Conditioning and Refrigeration Institute Standard ARI 700-88. *Note: ARI 700-88 sets the purity standard for reclaimed refrigerant and is also the standard for virgin refrigerant, Table 6-2. Many reclaim programs are being created that will allow the technician to send in or trade recovered refrigerant for reclaimed refrigerant.*

Table 6-2. ARI Standard 700-88, Purity for Reclaimed and Virgin Refrigerant

Contaminant	Allowable Limit by Weight
Moisture	10 ppm
Oil	100 ppm
Acidity	1.0 ppm
Chlorides	none

REFRIGERANT PURITY

The purity of reused refrigerant is a prime concern to the technician. If contaminated refrigerant is reintroduced to the system that has just been serviced, system failure will result.

Purity Field Test

The service technician performs field tests while working on a system. In field testing, the technician looks for evidence of moisture, acid, particulates and non-condensables.

Moisture. Most recycling machines have sight glasses that contain moisture indicators. These indicators change color when the moisture level is low enough for refrigerant reuse. A sight glass may also be installed when hooking up a recycling machine. Filter driers are designed to trap moisture that is present in the refrigerant.

Acid. Acid test kits can determine the level of potentially dangerous acids. Burnout filter cores can then be used to minimize the presence of identified acids.

Particulates. Good filtration systems have the capability of removing particulates from refrigerant.

Non-condensables. Many machines have a self-purging feature that will allow the purging of non-condensables. In addition, there is a field procedure that can be used to check for non-condensables (detailed below).

New products are being developed as the need for field testing increases. New field test kits that test for both acid and moisture content simply by taking a vapor sample of the recovered refrigerant are now on the market.

Field Test for Non-Condensables
This test should be performed on a storage cylinder stored at a temperature in excess of 65 °F. Perform the following steps when field testing for non-condensables:

1. Store the cylinder standing upright for 12 hours.
2. Install a calibrated compound gauge and determine the pressure in the standing cylinder.
3. Measure the air temperature next to the cylinder.
4. Using a temperature-pressure chart, determine the pressure corresponding to the air temperature for the refrigerant.
5. Compare the cylinder pressure and the pressure determined to correspond from the temperature-pressure chart.
6. If the pressure in the cylinder exceeds the corresponding air pressure by 10 psig (R-12), then non-condensables are present.
7. Vent a small amount of the container's refrigerant into the recycle machine and check pressures again.
8. If the pressures are now equal to or lower than the air-temperature-pressure relationship, the refrigerant no longer has non-condensables.
9. If the pressures are higher, then the refrigerant has non-condensables and must be recycled or reclaimed.

Laboratory Testing
ARI 700-88 standards are used when the refrigerant is tested in the laboratory. These standards establish an extremely high purity level for refrigerants. ARI 700-88 can only be achieved in a refrigerant reclamation process and not by most of the smaller recycling machines used in normal service. ARI Standard 700-88 is reprinted in its entirety in Appendix B.

THE RECYCLING PROCESS

A recycling machine recovers the refrigerant and cleans it. Recycling units come in a single- and dual-pass configuration. Chapter Seven will offer more details on single- and dual-pass machines, but for now, all we are concerned about is tracing the refrigerant flow through the machine as it is being cleaned. Figure 6-1 traces refrigerant flow through a composite machine and shows some of the features of a single-pass recycling unit. For a better understanding of the recycling process, read the following steps while tracing the refrigerant path in Figure 6-1:

1. Beginning at the box labeled recovery unit, the refrigerant enters the recycling unit.
2. The refrigerant passes through an oil separator and accumulator.

3. Oil that has been removed from the system is separated out at the oil separator. The drain under the separator allows for precise measurement of the oil removed from the system. System oil removed in recovery should be replaced with new oil.
4. The accumulator traps liquid refrigerant before it can damage the machine's compressor.
5. The refrigerant now passes through an acid core filter to remove acid that may exist in the refrigerant.
6. Refrigerant enters the compressor as low-pressure vapor and leaves as high-pressure vapor.
7. The high-pressure vapor passes through another oil separator.
8. The second oil separator removes oil that has been taken from the recovery machine's compressor and returns that oil to the recovery unit's compressor. Some units use a drain valve that will measure the oil so that new oil can be added.
9. The refrigerant passes through a filtration system. The filtration removes moisture and particulates.

Figure 6-1. Refrigerant Recycling

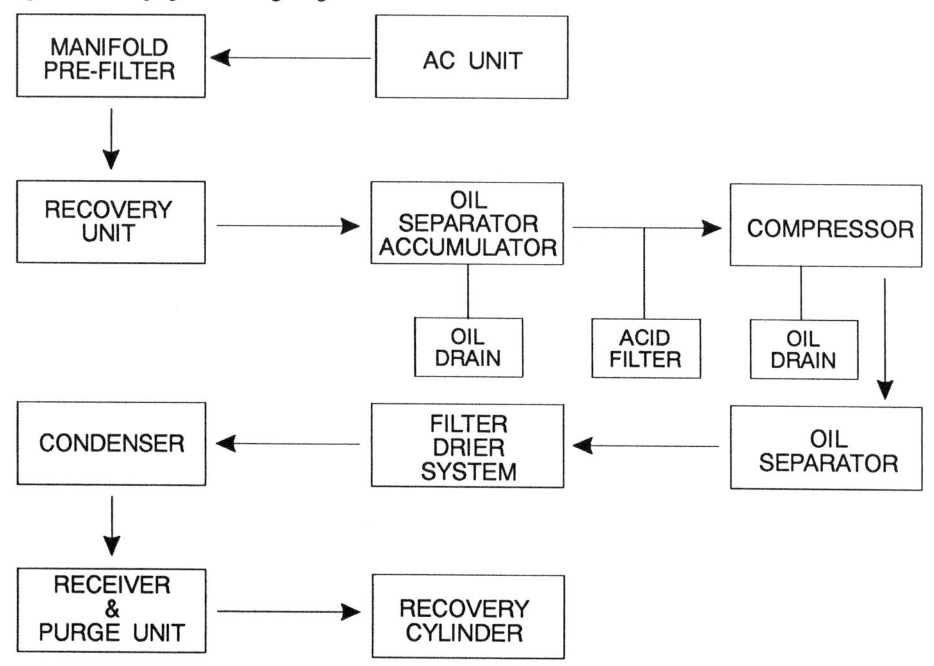

10. Entering the condenser, the refrigerant changes from high-pressure gas to high-pressure liquid.
11. The high-pressure liquid enters a receiver or cylinder that will hold a quantity of refrigerant. Non-condensables are purged automatically.
12. Some refrigerant remains in the receiver. In the case of certain machines, charging can take place from the receiver. The balance of the refrigerant flows to an external storage cylinder, completing a single-pass recovery/recycling process.

RECYCLING CONCERNS

Refrigerant purity and reuse raise concerns that only time and industry practice will ~~be~~ able to answer. Recycling as a procedure is still new. Many areas of accepted service procedure will undergo change to successfully service refrigeration units and at the same time comply with the CAA.

Consider for a moment some of the concerns that recycling raises:

1. *Does the use of recycled refrigerant void the manufacturer's warranty?* Many manufacturers insist that refrigerant returned to a system meet the ARI 700-88 standard. The standard cannot be reached on a rooftop since it requires a reclaim process to achieve. What happens, for example, if a new compressor is installed and fails thirty days after installation and the windings on the compressor are diagnosed as bad? Is the manufacturer of the new compressor liable under warranty? Or, is the recycled but possibly not totally acid free refrigerant the culprit? Was an acid test performed before the refrigerant was reused? Can the technician verify the results of that test?

2. *What should a customer be charged for recycled refrigerant?* It is the customer's refrigerant in the first place. Should a recycling charge become standard? Should you charge the same rate for recycled refrigerant and new, reclaimed or virgin refrigerant? What type of warranty should be offered on reclaimed and recycled refrigerant?

3. *How will using recycled or reclaimed refrigerant affect the customer's perception of your firm?* Customers are individuals, and some will think they are being taken advantage of if "used refrigerant" is used. Much has to be done in the way of education.

4. *What provisions will be established to ensure that recycling is really practiced by your competitor?* Are we entering an age of monitors

checking on compliance with the CAA? What type of penalties should be installed to punish noncompliance with the CAA? Will hvac/r personnel be certified to use recycling equipment?

These are just a few of the issues that the CAA raises, yet each of these issues can seriously impact the profitability of any contractor.

Hvac/r equipment is designed to use virgin refrigerant. Requirements of the CAA, refrigerant availability and cost are now creating a scenario wherein the technician must decide whether or not to recycle, and in doing so, use a refrigerant in less than its virgin state. This decision is not an easy one. Unfortunately at this stage, most of these concerns will be answered only by time and many completed service calls.

RECOVERY/RECYCLE IN THE AUTOMOTIVE INDUSTRY

Recovery/recycle in the automotive industry is different than recovery/recycle in hvac/r. The major difference lies in the quantity of refrigerant to be recycled, the purity of the recycled refrigerant, and the portability of the recycling units. Finally, the hvac/r industry uses many different refrigerants, and a recycling machine must possess multiple refrigerant handling capabilities.

Most of the recycling machines on the market have been developed to meet SAE standards and be UL (Underwriters Laboratories) approved. The SAE standards are not designed for hvac/r use and the UL approval does not establish purity standards for hvac/r use. The standards for recycling machines have been set and listed in ARI 740-1993. The EPA standards show the levels and standards that must be met depending on the type of service performed. Equipment purchased today must meet those standards. ARI also offers a list of approved equipment.

Chapter Seven will discuss ARI 740-91, which establishes some standards that may be applicable to the hvac/r industry.

Recycling in the automotive field does not create as many potential problems as it does in hvac/r. Automotive air conditioning systems are different than residential and light commercial systems. Automotive charges are smaller and the compressor electrical system is not involved with the system refrigerant (although this may change).

Burnout, not a problem in auto air, is a problem in hvac/r. Burnout seriously contaminates refrigerant. Returning refrigerant to system use after a burnout requires a major cleanup. Technology is available to achieve the required cleanup, but it takes time to clean up after a major burnout. In addition, the field testing for burnout must be thorough, or a new burnout will occur.

CONCLUSION

Recycling aids in reducing the ozone problem by stopping the venting of refrigerant into the atmosphere. In the future, less CFCs and HCFCs will have to be manufactured since the system under repair is receiving its own refrigerant back.

There are also economic benefits to the recycling process. Mandated production cutbacks and taxation are causing an escalation in the price of refrigerants. Recycling recovered refrigerant is one way to combat spiraling prices.

From the practical standpoint, recycling is a new process. Many new policies and procedures will have to be implemented to make recycling a workable system in the hvac/r field. These policies and standards should be established based on operating history in order for the CAA to accomplish its ultimate goal of reducing ozone depletion.

CHAPTER SEVEN

RECYCLING MACHINES

RECOVERY/RECYCLING MACHINES
Stand on any street corner and watch the variety of cars go by. Every car essentially has one purpose - to get the driver from one point to another. Yet, there are many makes and models of cars from which to choose. Each make and model has special features that make it different from its competitor. Recycling machines are no different than cars. The number of different makes and models of recycling machines having different features make choosing a machine confusing to say the least.

Many features in recycling machines are necessary to ensure acceptable performance. Other features, which some consider to be just bells and whistles, can make recycling easier, safer and can result in a purer refrigerant. Different manufacturers add or subtract different features. What is important is that the machine fulfills the needs of the user.

Recovery/recycling machine capabilities must be identified by the purchaser. Capability is measured in the speed of recovery and the ability of the machine to remove contaminants when recycling is needed. Need is measured by the purchaser's service requirements.

RECOVERY/RECYCLING/ RECLAIM STANDARDS
The CAA states that anyone working on automotive air conditioning had to use certified recycling equipment by January 1, 1992. By July 1, 1992, CFCs and HCFCs could not be vented to the atmosphere. This non-venting requirement makes refrigerant recovery mandatory. Once the refrigerant is recovered, it must generally be recycled or sent to a reclaimer before it can be reused.

In the last few years, many standards have been developed for recycling equipment. These standards can be considered a prelude to the ultimate certification process for machines, which is required under the CAA. Standards that are currently in place will become the platform for future recovery and recycling machines.

The following standards are used for recovery and recycling operation and equipment:

Society of Automotive Engineers
SAE Standard J1991: The standard for automotive refrigerant purity that is used by most recycling equipment manufacturers.

SAE Standard J1990: Extraction and recycle equipment for mobile air conditioning provides specification for R-12 recycling equipment and recovery equipment intended for automotive use.

Underwriters Laboratories
UL Standard 1963: Provides a basis to evaluate that automotive recovery/recycling equipment adequately recovers and recycles R-12. Refrigerants used in other applications are judged on standards in 1963 only as they apply. In the area of electrical and fire safety, UL 1963 for hvac/r does apply. UL 1963 does not concern itself with refrigerant purity for hvac/r applications. **Currently there are no accepted purity standards for the hvac/r industry when recycling is employed.**

Air-Conditioning & Refrigeration Institute
ARI 700-88: The purpose of this standard is to enable users to evaluate and accept/reject refrigerants regardless of source (new, reclaimed and/or repackaged) for use in new and existing refrigerating and air conditioning products within the scope of ARI. *Note: Reclaimed refrigerant must be tested in a laboratory to ensure that it meets the standards set forth in ARI 700-88. ARI 700-88 is reprinted in Appendix B.*
ARI 740-91: The purpose of this standard is to establish methods of testing for rating and evaluating performance of refrigerant recovery, recycle and/or reclaim equipment (herein referred to as equipment) for contaminant or purity levels, capacity, speed, and purge loss to minimize emissions into the atmosphere of designated refrigerants. 740-91 provides standards for equipment comparison, it does not establish the basis for acceptance of particular equipment. ARI has published a list of certified equipment. ARI 740-91 is reprinted in Appendix C.

American Society of Heating, Refrigerating, and Air-Conditioning Engineers, Inc. (ASHRAE)
ASHRAE Guideline 3-1990: Recommends practices and procedures that will reduce inadvertent release of CFCs in many areas of hvac/r manufacture and service. 3-1990 is designed to be used with standards and codes that are already in use.

PURCHASING CONSIDERATIONS

There are many different manufacturers of recovery and recycling equipment. It is impossible to recommend one machine over another; that is the responsibility of the buyer. The potential buyer must first analyze his or her future recovery/recycle requirements.

There is no sense in buying a machine that is too much or too little. Someone who services automobiles and operates a small business does not need a costly machine that involves a complicated filtration system designed to handle multiple refrigerants. In automotive service, the only refrigerant that is currently being recovered is R-12. Why pay for a machine to handle more than

R-12? The pay back potential is just not there. In the future, R-134a will have to be recycled, so consideration should be given to purchasing a machine that can handle R-134a. R-12 and R-134a can never be mixed, so a dedicated R-12 machine must be thoroughly cleaned before using R-134a.

Someone who services hvac/r equipment will need a machine that is capable of dealing with many different types of refrigerant. In addition, the physical size of the machine is a serious consider-ation. Many service calls involve carrying the machine up a ladder. Purchase a machine that has enough portability to meet your needs.

Recovery rate and the size of the systems normally serviced also need consideration. Machines on the market can handle systems holding charges up to 100 lbs, recover liquid at 3 lbs/min and vapor at 1/2 lb per minute. There are also machines that easily handle systems in the 100 to 1000 lb. range. Even larger machines exist that are designed to recover systems holding charges in excess of 1000 lbs. Match the machine to your needs - not to the salesperson's presentation.

Finally, cost versus available features should be considered. Every machine on the market has some special features. Do you need those features in your operation? If the machine is certified, meets your size and recovery needs, and possesses a good warranty, anything extra should be viewed with caution. It is safe to say that as time goes on, recovery/recycling technology will improve. Do not go overboard today, as tomorrow's machines will certainly make the whole recovery/recycle process easier.

RECYCLING EQUIPMENT COMPONENTS

Technology is moving forward in the machine market. Many weak-nesses have been identified and future machines will feature improvements as standard compo-nents. For example, today's ma-chine can stand improvement in the area of metering devices, from both a time and protection stand-point. Improvements are also needed in the changeover from one type of refrigerant to another. Finally, detection of non-condensables and purging detec-tion and operation will be improved in the future.

As with buying a car, next year's model will have new and improved features. The problem is that industry cannot wait years to buy a machine. The time clock for CAA compliance is running now.

Different machines have options that make for an easier recovery/recycling process. For certification purposes, many machines have certain operational features. The following list defines some of the components that are available in the marketplace today. The list does not deal with every possible component and is meant to give general descriptions of the components involved in recovery/recycling machines built today.

Single-and Dual-Pass Machines
The function of a recycling machine is to clean the refrigerant so it can be reused. The end result of recycling must be a clean refrigerant. Recycling equipment can be obtained in two basic configurations:

Single Pass: A single-pass recycling system will allow the technician to recover refrigerant and then clean the refrigerant through filtering and oil separation. The refrigerant makes a single pass through the recycling machine. After this pass, it is returned to the system or to a storage tank. Figure 7-1 is a composite drawing of a single-pass system.

Figure 7-1. Single-Pass Recycling

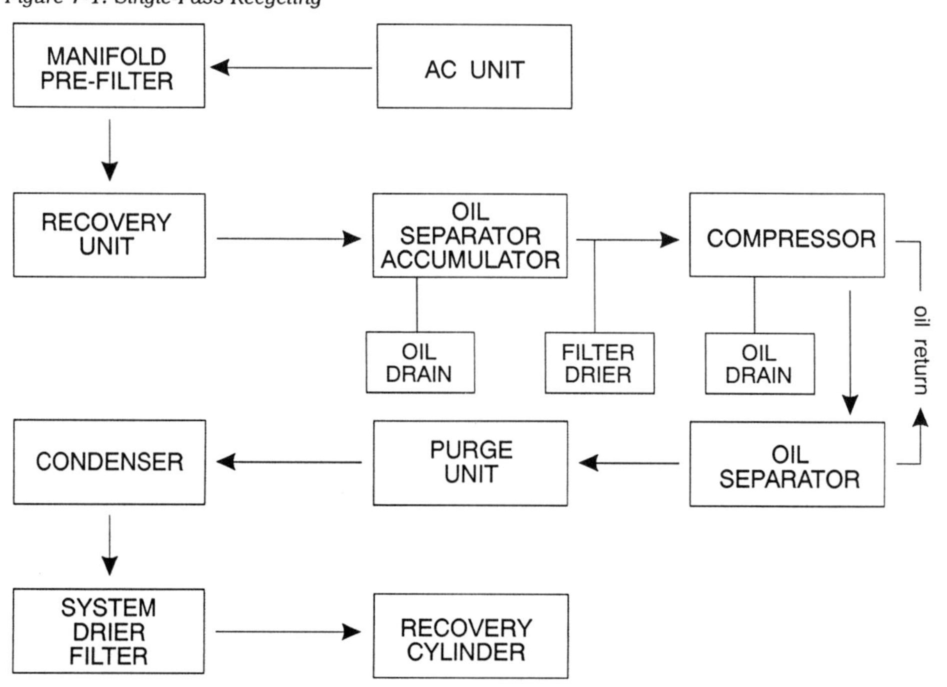

Dual Pass: The dual-pass system allows recovered refrigerant to be recycled in a single pass. At the operator's option, the refrigerant can then be recycled again. The additional recycling is accomplished by allowing the refrigerant to loop through the filter system as many times as are necessary to accomplish necessary cleaning and filtration.

Figure 7-2 is a composite drawing of a dual-pass machine. Neither Figures 7-1 nor 7-2 represent actual machines. Every machine will have its own specific operating system.

Figure 7-2. Dual-Pass Recycling

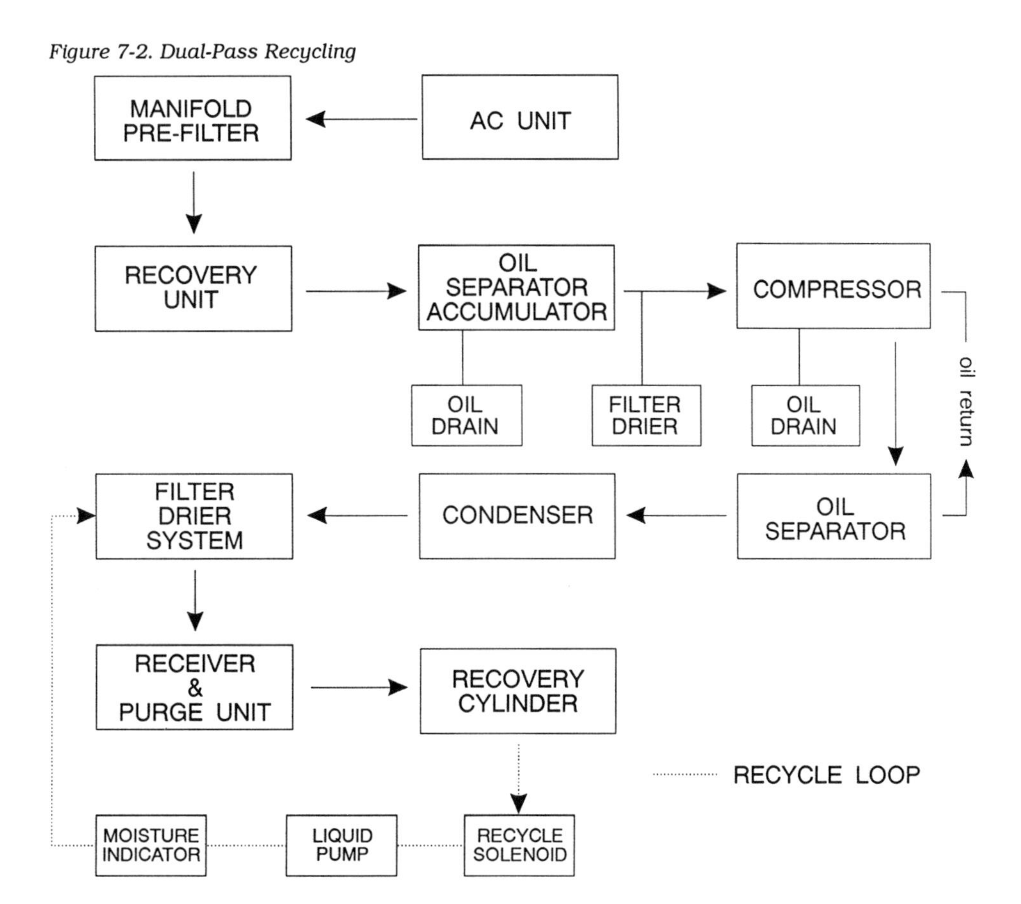

Accumulator
As in operational units, the accumulator serves to capture liquid refrigerant before the refrigerant can ruin the compressor valves in the recovery/recycle machine.

Oil Separator
The oil separator traps the oil that is removed from the system undergoing recovery. Some recovery/recycle machines contain an oil separator to capture oil that has been removed from the recovery/recycling machine's compressor and return that oil to the compressor.

Many machines have ports that allow the oil flow to be measured. This ensures that the correct amount of new oil is returned to the system under service or the compressor in the recovery/recycling machine. A combination oil separator and accumulator is also offered as an option by some manufacturers.

Oil is going to be a continuing problem in the new age of service. Oil that has been removed from a system should be replaced by new oil. This leaves the technician with a disposal problem - what to do with the old oil? In the current environment, refrigerant that is intended to be shipped to a reclaimer is not classified as a hazardous waste. This is not the case with refrigerant oil. Effectively, hazardous waste procedures will have to be observed when shipping oil. Some reclaimers will probably accept the used oil for disposal purposes, but arrangements will have to be made on an individual basis. Be sure to check the regulations in place when dealing with oil. **Under no circumstances can you dump the oil on the ground or in the trash can. Violators of this rule will be dealt with rather severely.**

Heat Exchanger
A heat exchanger is added to the oil separator to prevent liquid slugs from entering the compressor.

Liquid Pump
A liquid pump circulates refrigerant through the filtration system. The pump can be driven via magnetic coupling. Use of the pump eases wear and tear on the compressor and decreases pressure buildup and heat in the reclaimer/recycler.

Self-Purging Device
Non-condensables are a problem in any system. Manual and automatic purging features are available on recycling machines. Purging devices generally work by sensing higher than normal pressures, which indicate the presence of non-condensables.

High-Pressure Safety
There is more than one possible use for a high-pressure safety. One use is to act as a safety when pressures in the recycling machine or tank are too high. Automatic shutdown will then occur. It can also be set to signal the presence of non-condensables and alert the operator to their presence.

Low-Pressure Shutoff
Depending on the manufacturer, recovery/recycling machines may use a low-pressure shutoff, which turns the machine off at a set psi. Suction gas may cool the compressor on the recovery/recycle machine. As the low side pressures drop during recovery, there is a lack of cool suction gas to cool the compressor. The low-pressure shutoff is designed to protect the compressor by shutting down the recovery system.

Compressor Oil Sensing Unit
Some oil will always travel with refrigerant. The continuing loss of oil from a recovery/recycling machine compressor will cause compressor failure. The existence of an oil sensing unit will alert the operator to this possibility.

Filters
There are many different types of filters in use today. Filters can be used as a pre-filter connected outside of the machine, or acid filter kits can be purchased to remove acid before it enters the recovery unit. As to the actual filters in the recycling unit, different manufacturers use different mediums to filter, dry and trap. Which type is best is a matter of intended use and application within a machine. The recycling machine buyer should check for filter availability, since filters will have to be replaced on a regular basis. Cost for replacement filters and number of pounds of refrigerant that can be recycled without changing the filters should also be considerations in purchasing a machine. Consideration should also be given to whether or not the machine possesses the capability to advise the operator that the filters need changing. *Note: Since filters can and do trap oil, there is a problem regarding disposal of used filters. Check your EPA regulations, local regulations and reclaimers for the proper disposal procedures.*

Ability to Handle More Than One Type of Refrigerant
Some systems are designed to handle only R-12. In the future, there will be a need for the recovery equipment to handle the new alternative refrigerants. Hvac/r service requires the ability to handle multiple refrigerants. The machines should make the changeover from one type of refrigerant to another easily. This changeover may also involve changing filters, so ease of filter change is a valid consideration. Finally, you cannot recover R-12 from a system and pump it into a storage cylinder containing

previously recovered R-22. For this reason, extra tanks are a necessary requirement.

Oil Measurement
During refrigerant recovery, oil can be removed from the system. A calibrated method that allows the technician to remove the oil from the machine and measure how much has been removed ensures that the recovered oil is returned to the system in the proper amount. Measurement of oil quantity removed from a recycling machine compressor is also a consideration. Machines today have drain ports that allow oil recovered from a system or from a machine's compressor to be measured. This ensures that the proper amount of new oil is returned to the system or machine's compressor.

Portability
Expected usage is the major consideration in this category. The weight and size of the machine must be accounted for in rooftop service. Some firms may consider purchasing smaller, portable recovery units and a stationary recycling machine for the actual recycling process.

Machines are also on the market that separate the recovery function from the recycling function. After recovery is complete, then the recovery machine is attached to the recycling portion of the machine for refrigerant recycling. Do not count on running 100 feet of hose to the rooftop from the machine. Hose strength, ambient temperature, regulation and recovery time will make this practice a problem.

Approved Recovery Cylinders
An approved storage cylinder must always be used, regardless of whether the recovery/recycle mode is vapor or liquid. Valves on the storage tank should allow for both vapor and liquid recovery.

Current approved cylinders are DOT CFR TITLE 49, DOT 4BW, DOT 4BA or UL approved cylinders. In addition, the tank will carry a date stamp, which will determine when the cylinder must be re-certified for use. A BBQ tank is not an approved cylinder and none of the disposable refrigerant cylinders are legal or safe to use as recovery tanks. One of the major benefits of machine certification is that manufacturers will be selling recovery cylinders that are designed to do the job safely.

Reclaim processors and refrigerant manufacturers also have legal requirements that must be met for refrigerant storage. Check with your reclaimer and refrigerant supplier as to the legality of storing refrigerant in their cylinders.

Finally, the 80% maximum fill rule at a temperature of 70 °F must always be observed. Some type of protection should exist for shutting the machine down when recovery has reached the 80% mark in the storage cylinder. When the temperature increases above 70 °F, the room for expansion in the storage tank is decreased. **Always allowing more than 20% as a safety margin for refrigerant expansion can be a life-saving safety procedure. If the storage tank has some oil in it, there will be less room for refrigerant and this decrease must be compensated for.**

CHAPTER EIGHT

RECLAIMING

ASHRAE 3-1990 defines reclaim as "to reprocess refrigerant to new product specifications by means which may include distillation. Will require chemical analysis of the refrigerant to determine that appropriate product specifications are met. This term usually implies the use of processes or procedures available only at a reprocessing or manufacturing facility."

Reclaimers are now required to return refrigerant to the purity levels specified in ARI 700-1988 and to verify this purity level using the laboratory protocol set forth in the same standard. This is the standard set by the EPA for reclaiming.

DISTILLATION

Reclaiming refrigerant generally involves distilling the used refrigerant to ensure that any contaminants are removed. Reclaiming is performed by a chemical manufacturer or a dedicated reclamation plant. Once the reclaim process is finished, the refrigerant is laboratory tested to make sure it conforms to industry standards.

The distillation process involves heating the contaminated refrigerant to its boiling point, causing the refrigerant to vaporize. The vapor is piped into another cylinder, while the contaminants are left behind. These contaminants could not vaporize with the refrigerant, because precise boiling point temperature control allows only the refrigerant to change state. The vapor then condenses, resulting in pure refrigerant.

Reclaiming is not a field procedure. Even if you possessed a machine that could return the refrigerant to a virgin state, it would still be necessary to perform a chemical analysis. This chemical analysis must be performed in a laboratory environment. The laboratory looks for the following items in reclaimed refrigerant:

- Particulates/Solids
- Water
- Non-condensables
- Chloride ions
- Other Refrigerants
- Acidity
- High Boiling Residue

When the analysis is completed, the refrigerant complies with the ARI 700-88 standard and is certified to be pure enough to be reused. Reclaiming is necessary when recycling cannot return the refrigerant to a pure enough state to be reused.

Reclaiming is a process that is comparatively new in the refrigeration industry. Reclamation facilities exist throughout the United States, and the number of reclaimers is growing rapidly.

Projections indicate that existing reclaimers are also expanding into new geographic locations. As the ranks of reclaimers continue to swell, new methods for purchasing, handling, shipping and selling reclaimed refrigerant will be instituted. As the market grows, it should be just as easy to purchase reclaimed refrigerant as it is to buy a remanufactured part for your car.

WHY RECLAIM?

Reclaiming makes good business and environmental sense. Two factors that should influence the decision to use reclaimed refrigerant are cost and future supply.

Cost

Due to the ozone depletion taxes, the cost of refrigerant is skyrocketing. Reclaimed refrigerant is exempt from the ozone depletion tax, therefore its price is substantially lower than new refrigerant. However, while recycled refrigerant may appear cheaper than reclaimed refrigerant, this may not always be the case. There are hidden costs to recycling that will be covered later in this chapter.

Future Supply

Production cutbacks mandated under the Montreal Protocol will affect future refrigerant availability. Recent statistics show that production of R-12 is currently less than the allowable mandated production. In view of this, a shortage could take place sooner than expected. Industry estimates that 30% of the R-12 used in the future will come from existing systems.

Regulation

Under the new regulations, you cannot resell recycled refrigerant to another party. This may change as the EPA sets standards for recycled refrigerant purity. For now, you must reuse refrigerant in the system, or another system owned by the same person, or reclaim.

THE RECLAIMING PROCESS

The first step in the reclaiming process is to determine whether or not reclaiming is necessary. Assume that your customer is on a small system and your firm has been servicing the unit regularly. The service call involves a procedure that, by nature, ensures the technician the refrigerant in the system can be recovered in a usable state. Recovery and recycling would therefore be the recommended procedure. If the technician follows proper procedure and the recovery/recycling machine is clean, then recovery, recycling, and then returning the refrigerant to the system is all that is needed. Many service calls will fall into this category.

On the other hand, if the service call involves a suspected contaminated system, recycling is not the recommended procedure. Reclamation is necessary in these cases to avoid future problems.

Reclaiming Costs
If you handle large quantities of refrigerant, you have to know about refrigerant shipping, labeling and possible hazardous waste control procedures. Getting the refrigerant to the plant or to a distributor is part of the reclaim process and also part of the cost of reclaiming. Also, many processors demand a minimum quantity before they will perform the reclaim process.

Refrigerant sent to a reclaimer must first meet reclaimer standards. If those standards are not met, then the reclaimer must dispose of the refrigerant. The contractor must pay the shipping charges to get the refrigerant to the reclaimer, then pay a disposal handling fee if the reclaimer refuses the refrigerant. If this is the case, the contractor pays both shipping and disposal charges and ends up with no usable refrigerant. *It is important to understand that the refrigerant submitted for reclamation is not automatically accepted by the reclaimer. The reclaimer considers the refrigerant to belong to the contractor until it is decided the refrigerant is reclaimable.*

If the refrigerant is suspected to have come in contact with a hazardous waste, a pretest should be considered. Pretesting involves sending a sample of the suspected refrigerant to the reclaimer for testing to determine whether or not the refrigerant can be reclaimed. If recovered refrigerant comes in contact with a hazardous waste, the reclaimer will not accept that refrigerant for an ordinary reclaim procedure. Pretesting will identify the presence of hazardous wastes and arrangements can then be made to dispose of the refrigerant locally. This saves the contractor money.

Paperwork
Paperwork is always necessary when shipping refrigerant to a reclaimer. The following documentation must accompany the refrigerant:

Ownership. It is not the reclaimer's refrigerant until the reclaimer accepts ownership.

Suspected contaminants. The contractor should identify any suspected contaminants in the refrigerant.

Proper labels. The DOT requires that all refrigerant being shipped bear a label properly identifying the refrigerant.

Shipping. Since refrigerant cannot be shipped in just any container, the reclaimer may provide the steel drums used in recovery and subsequent shipping. Arrangements must be made with the reclaimer to legally use his or her containers for shipping. In many cases, you cannot ship refrigerant back to a reclaimer in the same cylinder you received virgin refrigerant. Each reclaimer will have a policy stating filling requirements and container usage, because the container requirements for low-pressure refrigerants may differ from high-pressure refrigerants.

REFRIGERANT DISPOSAL
Refrigerant disposal is an area that will become more complicated. You cannot throw used refrigerant in the dump; it must be handled at a waste disposal facility. Some states have already established hazardous waste requirements for refrigerant, although the EPA does not consider used refrigerant to be hazardous waste. Check your state for its particular requirement.

Old filter cores and oil are accepted by some reclaimers. The shipping of these materials can be a problem, though, because it may be considered hazardous waste.

NEW MARKETING APPROACHES
As the demand for reclaimed refrigerants grows, the marketing of reclaimed refrigerant will change. One current method involves the contractor selling used refrigerant to the reclaimer for so much per pound. The price paid by the reclaimer can be applied as a credit toward the purchase of virgin refrigerant, or it can be taken in cash.

It is important to note that since reclaiming is relatively new, there will be many changes in its future. If you do not subscribe to refrigeration industry publications or belong to refrigeration associations, now is the time to start. Remember, it is the informed contractor who gains the edge in the marketplace.

RECYCLE VERSUS RECLAIM
When a system holds a small, uncontaminated charge, recovery and recycling are viable options. However, there are many instances when the only choice will be to use reclaimed refrigerants. Recycled refrigerant looks more attractive because, on the surface, it doesn't cost anything, while reclaimed refrigerant costs so much per pound. Since the contractor has recycling equipment on site, why not recycle? Recycled

refrigerant costs less than reclaimed refrigerant, so why reclaim? The truth is, there are factors that should be considered when trying to decide whether to use recycled or reclaimed refrigerant. These factors are the hidden costs associated with recycled refrigerant.

Recycling Machine Maintenance

Recycling machines are expensive, so it is smart to keep the machines properly maintained. This includes changing the filters at specified intervals. Filters may also have to be changed when refrigerants are mixed or when recovered refrigerant is severely contaminated. If pre-filtering is employed, then there is the additional expense of replacing the pre-filter.

Time

It takes time to recycle refrigerant, and time is money. If the refrigerant is dirty, many passes through the machine will be needed. And, filters can take up to 72 hours to do a thorough job. According to SAE J1989, if non-condensables are suspected, the refrigerant must stand for 12 hours before a final check can be made. Who pays for the time spent in rechecking the refrigerant after the 12 hours expire? These particular concerns do not exist when reclaimed refrigerant is used.

Purity

When recycled refrigerant is used, there is no effective method of determining purity. Prior service calls could mean the refrigerant could be mixed, or contaminated refrigerant could have been introduced into the system. Also, if a technician uses recycled refrigerant that contains acid, a system with a hermetic compressor will fail. The acid present in the refrigerant will attack the motor windings, causing compressor failure. Failure results in a costly repair to the unit. Reclaimed refrigerant, on the other hand, is always certified, so its purity standards are equivalent to virgin refrigerant.

Hazardous Waste Handling

Currently, used refrigerant is exempt from shipping and handling requirements that apply to hazardous waste materials. Certain states have legislation stating used refrigerant is hazardous waste, while others are in the process. Handling, shipping and disposal costs for hazardous waste are very high. In addition, contaminated refrigerant oil and filter cores may be considered hazardous waste materials.

Purging Air and CFC Release

Refrigeration systems cannot tolerate air, but purging a system to eliminate air can result in refrigerant discharge. This discharge might be considered accidental, but if proper procedures and equipment are not employed, the technician is at risk.

Mixed Refrigerants
Reclamation plants are finding that a significant amount of refrigerant submitted for reclaiming is mixed. Whether this mixing occurs because of improper recovery operation or poor service procedures has yet to be determined. As new alternative refrigerants come on line, though, mixing will become an even greater problem if recycling is employed. The alternative refrigerants cannot be mixed, and in some cases, a mixture of refrigerants reintroduced into a system can cause system damage. Reclaimed refrigerant solves the mixing problem, because mixed refrigerants can be separated out in the reclaiming process. However, in some cases, mixed refrigerants will not be reclaimed. The chemical analysis required before reclaiming can take place usually detects the presence of a mixture. When reclaimed refrigerant is used, mixed refrigerant is not a concern.

Oil and Associated Problems
Oil is removed with refrigerant during the recovery process, and this oil must be replaced, or lubrication problems will develop in the system. Besides this potential problem, some of the oils used in the new alternative refrigerants are not compatible with chlorine. If a recycling machine is not properly cleaned, it may contain chlorine. The introduction of this residual chlorine with the oil used in some alternative refrigerants will result in the chlorine forming an acid.

Lack of Analysis
When all factors are considered, tomorrow's technician will have to make difficult choices on a daily basis. Any decision requires the proper information to be correct. The lack of analysis of the recovered refrigerant will prove to be the biggest problem facing our industry. There are field procedures in place to determine the presence of non-condensables, acid and moisture. These field procedures are not detailed enough to gauge the purity levels necessary for every system's proper operation, though. There is no method outside of analysis that can certify the refrigerant being returned to the system is not contaminated. When in doubt, consider reclaiming.

RECLAIMED REFRIGERANT PURITY STANDARDS

Appendix B contains ARI 700-88, which sets the standard for reclaimed and virgin refrigerant. These standards can only be verified in the laboratory environment. Just because your recovery machine has a label that calls it a reclaimer does not mean it is a reclaimer. Refrigerant must meet ARI 700-88 to be considered reclaimed.

CHAPTER NINE

ALTERNATIVE REFRIGERANTS

The future of the refrigeration industry depends upon the acceptance and use of alternative refrigerants. It is a good thing that the use of alternative refrigerants will not occur all at once, since we still have much to learn about these alternatives. As CFC and HCFC production decreases, replacement refrigerant will come either from recycling, reclaiming or the use of alternative refrigerants.

SINGLE MIXTURES, AZEOTROPES, AND BLENDS

Single mixture, azeotrope and blend are three terms that should be understood by tomorrow's technician. The new alternatives are concentrating on single mixture replacements but as concerns develop in specific applications, blends are also being considered as potential alternatives.

Single Mixtures

R-12 and R-22 are common examples of single mixture refrigerants. These refrigerants are man-made compounds created to contain all the properties necessary to obtain refrigeration in a single mixture. New alternatives such as R-123 or R-134a are single mixtures.

Azeotropes

An azeotrope is created by mixing certain single mixture refrigerants. For example, R-502 is composed of 48.8% R-22 and 51.2% R-115. An azeotrope is created by the refrigerant manufacturer, not mixed in a bucket in the field. The refrigerants do not combine chemically, but they do provide the qualities necessary to perform as a single refrigerant.

Blends

Blends are relatively new. They can be described as many refrigerants working together to closely approximate the original refrigerant characteristics. Blends are created to meet specific operating needs. For example, a blend currently under development consists of 36% R-22 with 24% R-152A and 40% R-124.

Blends may present a new type of operating problem. In the event of a leak, the refrigerants comprising the blend will leak at different pressures, since their operating curves are different. Should a leak occur, a system starting out with a blend could theoretically end up with only one of the three original refrigerants remaining in the system.

Figure 9-1 shows the current alternatives that are either available or being considered as replacements for the existing refrigerants.

Figure 9-1. Alternative Refrigerants

Existing Refrigerants	Alternative Refrigerants	Other Alternatives
R-11	R-123	
R-12	R-134a	Blends or R-22
R- 22	R-22	
R-502	R-125	Blends or R-22
R-114	R-124	
R-500	R-134a	Blend

1. Alternative refrigerants should not be considered drop in replacements.
2. Alternatives will not replace in all cases

There are concerns about the new alternative refrigerants. These concerns are real, and they vary in degree, depending on both the refrigerant itself and the intended application. Some of these concerns are:

- Vapor pressures do not exactly match the original refrigerant's operating pressures.
- Existing oils, in some cases, will not work with the alternatives. Lubrication can become a problem in other cases.
- Toxicity levels in some of the new refrigerants are higher than those in the "old" refrigerants.
- It is almost always impossible to mix the original refrigerant and the alternative replacement. If R-12 is left in a recycling machine and R-134a is subsequently recovered, the system will be a mess.
- Alternative refrigerants cannot be used as drop-in replacements, as this will destroy the integrity of the existing system.
- Retrofitting existing equipment may not be cost effective when compared to the remaining useful life of the equipment. In many cases, components will have to be changed in order for the alternative to work. (Before any conversions are made, especially when chillers are involved, the original equipment manufacturer should be consulted.) It is senseless to convert a unit when the overall conversion cost compared to remaining equipment life does not justify pay back.
- Refrigeration systems operate under every conceivable condition, which can create more problems. A commercial application in Minneapolis, Minnesota operates in a totally different environment than the same piece of equipment operating in Phoenix, AZ. Operating under severely different conditions creates the need for different approaches when using an alternative refrigerant, and not all of these approaches have been discovered. There is very little equipment operating with the new alternatives, so the majority of our experience is limited to laboratory experiments.

THE ALTERNATIVES

Three new alternative refrigerants that are becoming available are R-123 (an HCFC), R-134a (an HFC) and R-22 (an HCFC). These refrigerants were developed as replacements for commercial and industrial chillers, medium-temperature air conditioning and refrigeration equipment.

The operating pressures of R-123 and R-134a are similar to the refrigerants they are replacing, but their performance levels are different, Figure 9-2. System capacity is a necessary requirement for a successful refrigeration cycle, and capacity is affected by operating environment and required evaporator temperature. The new alternatives will affect capacity. Capacity decrease will depend on the operating temperature of the unit being retrofitted and will generally occur with some equipment when the new alternatives are used.

Figure 9-2. Alternative vs. Existing Refrigerants, Temperature-Pressure

TEMP F	R-123	R-11	R-134A	R-12
-50	29.2in	28.9in	18.6in	15.4in
-40	28.8in	28.4in	14.7in	11.0in
-30	28.3in	27.8in	9.7in	5.5in
-20	27.7in	27.0in	3.6in	0.6
-10	26.9in	26.0in	2.0	4.5
-5	26.4in	25.4in	4.1	6.7
0	25.8in	24.7in	6.5	9.2
5	25.2in	23.9in	9.1	11.8
10	24.5in	23.1in	12.0	14.7
15	23.7in	22.1in	15.1	17.7
20	22.8in	21.1in	18.4	21.1
25	21.8in	19.9in	22.1	24.6
30	20.7in	18.6in	26.1	28.5
35	19.5in	17.1in	30.4	32.6
40	18.1in	15.6in	35.0	37
45	16.6in	13.8in	40.0	41.7
50	15.0in	12.0in	45.4	46.7
55	13.1in	9.9in	51.2	52.1
60	11.2in	7.7in	57.4	57.8
65	9.0in	5.2in	64.0	63.8
70	6.6in	2.6in	71.1	70.2
75	4.1in	0.1	78.6	77.0
80	1.3in	1.6	86.7	84.2
85	0.9	3.3	95.2	91.7
90	2.5	5.0	104.3	99.7
95	4.2	6.9	113.9	108.2
100	6.1	8.9	124.1	117.0
105	8.1	11.1	134.9	126.4
110	10.2	13.4	146.3	136.2
115	12.6	15.9	158.4	146.5
120	15.0	18.5	171.1	157.3
125	17.7	21.3	184.5	168.6
130	20.5	24.3	198.7	180.5
135	23.5	27.4	213.5	192.9
140	26.7	30.8	229.2	205.9
145	30.2	34.3	245.6	219.5
150	33.8	38.1	262.8	233.7

in = inches of mercury vacuum
Pressure as psig

R-123

R-11 is one of the first refrigerants targeted for replacement. There are over 80,000 chillers operating in the United States, many of which use R-11. Since the average chiller has a service life of 30 years, R-11 phaseout could seriously impair chiller capability in the United States. R-123 was designed to replace R-11 in order to limit this potential impairment. Estimates indicate that half of the chillers currently operating in the U.S. can be retrofitted to use an alternative refrigerant.

There is a big difference between R-123 and R-11. R-123 has an ozone depletion potential (ODP) of 0.02, and R-11 has an ODP of 1.0. Using R-123 lowers the ozone depletion potential almost 98%.

When a chiller is retrofitted to R-123, there is a loss of capacity and a decrease in cycle efficiency. Current tests show a capacity loss of 10 to 15%, and a cycle efficiency decrease of 2 to 5%. The rate at which capacity decreases is governed by the particular system. Higher speed impellers and larger systems will experience a greater system loss. Machines with slightly oversized compressors and expansion systems will experience a smaller loss.

When retrofitting a chiller, certain components may have to be changed. HCFC is a stronger solvent than R-11. Because of this, neoprenes, seals, motor insulation, gasket bushings, and diaphragms may need to be changed in order to be more compatible with R-123. Finally, different desiccants and resetting relief devices will be required.

Retrofitting a chiller to R-123 involves many factors. First and foremost is the expected life of the chiller. If the expected life is short and many components have to be changed out, it does not make economic sense to make the changeover from R-11 to R-123. In this case, it is better to purchase a new chiller that will operate on R-123.

Another consideration is the possible exposure to R-123. A monitor should be installed to alert the user to the presence of HCFC in the environment. Current exposure limits for R-123 are 10 parts per million (ppm) for an 8 to 12 hour day. However, there is an ongoing debate between some refrigerant manufacturers as to what the acceptable daily exposure level should be.

Other considerations are the exhaust devices (venting outdoors), air monitors and alarms that should be installed to limit exposure in the event of a leak. It is recommended that a portable breathing apparatus be maintained on site to be used for emergency purposes. Purge units to control random discharge should also be considered, as they are helpful in reducing emissions. A purge unit can reduce emissions up to 90% in a single installation. If an R-123 spill takes place, then the current allowable exposure is 1,000 ppm for one hour. This is the same exposure limit as R-11. Like most refrigerants, the biggest danger is asphyxiation. The space should be ventilated and an evacuation plan should be in place in case of leaks or spills.

R-134a

R-134a is the current favored replacement for R-12 and R-500 in certain applications. The development of R-134a has been increased to help relieve the potential R-12 shortage. The ozone depletion potential of R-134a is zero, because R-134a has no chlorine molecule.

Chillers operating in the medium-pressure range using R-12 and R-502 are prime targets for R-134a conversion. However, when a chiller using R-12 is converted to R-134a, compressor speed may also have to be increased.

One of the greatest contributors to the ozone problem is the automobile. The construction of automobile air conditioning is often predisposed to leaks, and leaks create the potential for ozone depletion. R-134a will become the prime refrigerant for use in automotive applications. Manufacturers are already marketing systems that use R-134a instead of R-12.

On the commercial side, there are already supermarkets in existence using R-134a as the sole refrigerant. The potential for R-134a to fully replace R-12 is still being studied, but equipment manufacturers are already working on the new generation of R-134a equipment.

R-134a may not be the final solution to the alternative issue. And, as with any refrigerant, there are problems with R-134a. One of these problems is oil. R-134a is used with a polyalkylene glycol (PAG) oil. PAG oil is not the same as the mineral oil used in R-12 applications. Mineral oil is insoluble with R-134a, so R-134a cannot be used as a drop-in replacement for R-12. In this case, it is not the refrigerant mixing that necessarily creates the problem, it is the oil mixing. PAG oils have also been shown to have some adverse effects on motor insulation and some of the polymers used in compressors.

Future service procedures will require dedicated recovery equipment for alternative refrigerants. A small amount of R-134a with a PAG oil introduced into a R-12 system with mineral oil will create havoc. Also, R-134a breaks down the desiccants used in R-12 systems, so the desiccants must be different in R-134a and R-12 systems. When this desiccant breakdown occurs, harmful acids form.

Figure 9-3 shows some of the operating characteristics of the alternative refrigerants as compared to current refrigerants. *Note: When exposed to high temperatures, the new alternatives can decompose. Compounds such as hydrogen chloride or hydrogen fluoride can be created. These compounds are acidic and pungent, causing irritation of the nose and throat. Immediate evacuation and medical treatment for exposure should be standard operating procedure.*

Figure 9-3. Alternative vs. Existing Refrigerants Performance Data, 5 °F Evaporating / 86 °F Condensing

	R-11	R-123	R-12	R-134a
Evaporating Pressure	23.9in	25.2in	11.8	9.104
Condensing Pressure	3.62	1.20	93.3	97.004
Compression Ratio	6.24	6.91	4.07	4.69
Compressor Discharge Temperature	104	91	100	97
Temperature Suction Gas	5	5	5	5
Specific Volume Suction Vapor cu. ft/lb	12.2	13.93	1.46	1.93
Latent Heat of Vapor Btu/lb	83.5	79	68.2	89.3
Net Refrigeration Effect Btu/lb	67.2	61	50.1	63.1
Coefficient of Performance	5.09	4.93	4.69	4.60
Horsepower / Ton of Refrigeration	0.926	.956	1.005	1.02
Refrigerant Circulated per Ton lbs/min	2.98	3.28	3.99	3.17
Comp. Suction Gas Volume per Ton cu. in/min	36.4	45.7	5.84	6.1
Liquid Circulated per Ton cu. in/min	56.3	62.9	85.5	73.90

in = inches of hg

R- 22
It may appear strange to see an HCFC listed as an alternative refrigerant, but it is a possible choice. In low- and medium-temperature applications, the use of R-12 and R-502 is being phased out, and the current replacement in many cases is R-22.

When R-22 was used in low-temperature applications years ago, there were problems. One of these problems was an excessively high discharge temperature when a unit was operating at low suction pressures. Today, this problem can be overcome to a certain degree by using two-stage compressors, compound compressors, or alternative methods of compressor cooling. All of these methods are expensive, but so are the alternatives. The demand for R-22 in low-temperature applications is growing. One manufacturer noted that orders for new equipment using R-12 or R- 502 only constitute 10% of total sales.

One major drawback to R-22 is that it retains more water at higher temperatures than R-12 or R-502. This means that as the temperature drops, the refrigerant retains more water. The same is true in its liquid state. To ensure a dry refrigerant, it is necessary to be very careful when performing evacuation procedures. In addition, brazing and manufacturing techniques must make sure that no cavities are created that will act as hiding places for water molecules.

Future Alternatives
Because of the problems that occurred when R-22 was used in low-temperature applications, R-502 was designed. There is a potential replacement for R-502 on the drawing board, but that replacement (R-125) is not yet available.

Refrigerant manufacturers are trying to compress decades of research into days in order to prepare for future needs. As the use of alternatives increases, the results will be better understood. Questions still exist, such as:

- Does the alternative perform optimally in the heat and cold?
- How does the refrigerant affect human health?
- What safety concerns do the new refrigerants create?

All of these concerns are being addressed in laboratories throughout the world. They are, however, questions that only time can answer. While conditions can be simulated in a laboratory, the only true test of the new refrigerants will be in the field, operating under every extreme for an extended period of time. It is tomorrow's technician who will write the final chapter on alternative refrigerants.

CHAPTER TEN

SAE STANDARDS

Automobile air conditioning is recognized as one of the major causes of ozone depletion. Automobiles leak refrigerant at a steady pace, and past service procedures consisted of adding to or replacing the charge on a regular basis. This practice resulted in a considerable amount of unnecessary chlorine being released to the atmosphere.

In October 1989, the Society of Automotive Engineers issued three standards that have had a significant impact on the ozone issue. These standards are cited in the Clean Air Act and constitute the partial groundwork for future automobile air conditioning service. One of the standards also provides the specifications for R-12 recycling and/or recovery equipment used to service automobiles.

It is interesting to note that although the standards apply to automotive air conditioning service, many of the service procedures flow well into the hvac/r industry.

THE FOLLOWING STANDARDS ARE REPRODUCED IN THEIR ENTIRETY:

1. SAE J1989 *Recommended Service Procedure for the Containment of R-12*

2. SAE J1990 *Extraction and Recycle Equipment for Mobile Air Conditioning Service*

3. SAE J1991 *Standards for Purity for Use in Mobile Air-Conditioning Systems*

These reports are published by SAE to advance the state of technical and engineering sciences. The use of these reports is entirely voluntary, and their applicability and suitability for any particular use, including any patent infringement arising therefrom, is the sole responsibility of the user.

400 COMMONWEALTH DRIVE, WARRENDALE, PA 15096

HIGHWAY VEHICLE RECOMMENDED PRACTICE

Submitted for recognition as an American National Standard

SAE J1989

Issued October 1989

RECOMMENDED SERVICE PROCEDURE FOR THE CONTAINMENT OF R-12

1. SCOPE:

During service of mobile air-conditioning systems, containment of the refrigerant is important. This procedure provides service guidelines for technicians when repairing vehicles and operating equipment defined in SAE J1990.

2. REFERENCES:

SAE J1990, Extraction and Recycle Equipment for Mobile Automotive Air-Conditioning Systems

3. REFRIGERANT RECOVERY PROCEDURE:

3.1 Connect the recovery unit service hoses, which shall have shutoff valves within 12 in (30 cm) of the service ends, to the vehicle air-conditioning system service ports.

3.2 Operate the recovery equipment as covered by the equipment manufacturers recommended procedure.

3.2.1 Start the recovery process and remove the refrigerant from the vehicle AC system. Operate the recovery unit until the vehicle system has been reduced from a pressure to a vacuum. With the recovery unit shut off for at least 5 min, determine that there is no refrigerant remaining in the vehicle AC system. If the vehicle system has pressure, additional recovery operation is required to remove the remaining refrigerant. Repeat the operation until the vehicle AC system vacuum level remains stable for 2 min.

3.3 Close the valves in the service lines and then remove the service lines from the vehicle system. Proceed with the repair/service. If the recovery equipment has automatic closing valves, be sure they are properly operating.

4. SERVICE WITH MANIFOLD GAGE SET:

4.1 Service hoses must have shutoff valves in the high, low, and center service hoses within 12 in (30 cm) of the service ends. Valves must be closed prior to hose removal from the air-conditioning system. This will reduce the volume of refrigerant contained in the service hose that would otherwise be vented to atmosphere.

4.2 During all service operations, the valves should be closed until connected to the vehicle air-conditioning system or the charging source to avoid introduction of air and to contain the refrigerant rather than vent open to atmosphere.

4.3 When the manifold gage set is disconnected from the air-conditioning system or when the center hose is moved to another device which cannot accept refrigerant pressure, the gage set hoses should first be attached to the reclaim equipment to recover the refrigerant from the hoses.

5. RECYCLED REFRIGERANT CHECKING PROCEDURE FOR STORED PORTABLE AUXILIARY CONTAINER:

5.1 To determine if the recycled refrigerant container has excess noncondensable gases (air), the container must be stored at a temperature of 65°F (18.3°C) or above for a period of time, 12 h, protected from direct sun.

5.2 Install a calibrated pressure gage, with 1 psig divisions (0.07 kg), to the container and determine the container pressure.

5.3 With a calibrated thermometer, measure the air temperature within 4 in (10 cm) of the container surface.

5.4 Compare the observed container pressure and air temperature to determine if the container exceeds the pressure limits found on Table 1, e.g., air temperature 70°F (21°C) pressure must not exceed 80 psig (5.62 kg/cm^2).

TABLE 1

TEMP°F	PSIG	TEMP°F	PSIG	TEMP°F	PSIG	TEMP°F	PSIG	TEMP°F	PSIG
65	74	75	87	85	102	95	118	105	136
66	75	76	88	86	103	96	120	106	138
67	76	77	90	87	105	97	122	107	140
68	78	78	92	88	107	98	124	108	142
69	79	79	94	89	108	99	125	109	144
70	80	80	96	90	110	100	127	110	146
71	82	81	98	91	111	101	129	111	148
72	83	82	99	92	113	102	130	112	150
73	84	83	100	93	115	103	132	113	152
74	86	84	101	94	116	104	134	114	154

TABLE 1 (Metric)

TEMP°C	PRES	TEMP°C	PRES	TEMP°C	PRES	TEMP°C	PRES	TEMP°C	PRES
18.3	5.20	23.9	6.11	29.4	7.17	35.0	8.29	40.5	9.56
18.8	5.27	24.4	6.18	30.0	7.24	35.5	8.43	41.1	9.70
19.4	5.34	25.0	6.32	30.5	7.38	36.1	8.57	41.6	9.84
20.0	5.48	25.5	6.46	31.1	7.52	36.6	8.71	42.2	9.98
20.5	5.55	26.1	6.60	31.6	7.59	37.2	8.78	42.7	10.12
21.1	5.62	26.6	6.74	32.2	7.73	37.7	8.92	43.3	10.26
21.6	5.76	27.2	6.88	32.7	7.80	38.3	9.06	43.9	10.40
22.2	5.83	27.7	6.95	33.3	7.94	38.8	9.13	44.4	10.54
22.7	5.90	28.3	7.03	33.9	8.08	39.4	9.27	45.0	10.68
23.3	6.04	28.9	7.10	34.4	8.15	40.0	9.42	45.5	10.82

PRES kg/sq cm

5.5 If the container pressure is less than the Table 1 values and has been recycled, limits of noncondensable gases (air) have not been exceeded and the refrigerant may be used.

5.6 If the pressure is greater than the range and the container contains recycled material, slowly vent from the top of the container a small amount of vapor into the recycle equipment until the pressure is less than the pressure shown on Table 1.

5.7 If the container still exceeds the pressure shown on Table 1, the entire contents of the container shall be recycled.

6. CONTAINERS FOR STORAGE OF RECYCLED REFRIGERANT:

6.1 Recycled refrigerant should not be salvaged or stored in disposable refrigerant containers. This is the type of container in which virgin refrigerant is sold. Use only DOT CFR Title 49 or UL approved storage containers for recycled refrigerant.

6.2 Any container of recycled refrigerant that has been stored or transferred must be checked prior to use as defined in Section 5.

7. TRANSFER OF RECYCLED REFRIGERANT:

7.1 When external portable containers are used for transfer, the container must be evacuated to at least 27 in of vacuum (75 mm Hg absolute pressure) prior to transfer of the recycled refrigerant. External portable containers must meet DOT and UL standards.

7.2 To prevent on-site overfilling when transferring to external containers, the safe filling level must be controlled by weight and must not exceed 60% of container gross weight rating.

8. DISPOSAL OF EMPTY/NEAR EMPTY CONTAINERS:

8.1 Since all the refrigerant may not be removed from disposable refrigerant containers during normal system charging procedures, empty/near empty container contents should be reclaimed prior to disposal of the container.

8.2 Attach the container to the recovery unit and remove the remaining refrigerant. When the container has been reduced from a pressure to a vacuum, the container valve can be closed. The container should be marked empty and is ready for disposal.

RATIONALE:

Not applicable.

RELATIONSHIP OF SAE STANDARD TO ISO STANDARD:

Not applicable.

REFERENCE SECTION:

SAE J1990, Extraction and Recycle Equipment for Mobile Automotive Air-Conditioning Systems

APPLICATION:

During service of mobile air-conditioning systems, containment of the refrigerant is important. This procedure provides service guidelines for technicians when repairing vehicles and operating equipment defined in SAE J1990.

COMMITTEE COMPOSITION:

DEVELOPED BY THE SAE DEFROST AND INTERIOR CLIMATE CONTROL STANDARDS COMMITTEE:

W. J. Atkinson, Sun Test Engineering, Paradise Valley, AZ - Chairman
J. J. Amin, Union Lake, MI
H. S. Andersson, Saab Scania, Sweden
P. E. Anglin, ITT Higbie Mfg. Co., Rochester, MI
R. W. Bishop, GMC, Lockport, NY
D. Hawks, General Motors Corporation, Pontiac, MI
J. J. Hernandez, NAVISTAR, Ft. Wayne, IN
H. Kaltner, Volkswagen AG, Germany, Federal Republic
D. F. Last, GMC, Troy, MI
D. E. Linn, Volkswagen of America, Warren, MI
J. H. McCorkel, Freightliner Corp., Charlotte, NC
C. J. McLachlan, Livonia, MI
H. L. Miner, Climate Control Inc., Decatur, IL
R. J. Niemiec, General Motors Corp., Pontiac, MI
N. Novak, Chrysler Corp., Detroit, MI
S. Oulouhojian, Mobile Air Conditioning Society, Upper Darby, PA
J. Phillips, Air International, Australia
R. H. Proctor, Murray Corp., Cockeysville, MD
G. Rolling, Behr America Inc., Ft. Worth, TX
C. D. Sweet, Signet Systems Inc., Harrodsburg, KY
J. P. Telesz, General Motors Corp., Lockport, NY

400 COMMONWEALTH DRIVE, WARRENDALE, PA 15096

HIGHWAY VEHICLE RECOMMENDED PRACTICE

Submitted for recognition as an American National Standard

SAE J1990

Issued October 1989

EXTRACTION AND RECYCLE EQUIPMENT FOR MOBILE AUTOMOTIVE AIR-CONDITIONING SYSTEMS

FOREWORD: Due to the CFC's damaging effect on the ozone layer, recycle of CFC-12 (R-12) used in mobile air-conditioning systems is required to replace system venting during normal service operations. Establishing recycle specifications for R-12 will assure that system operation with recycled R-12 will provide the same level of performance as new refrigerant.

Extensive field testing with the EPA and the auto industry indicates that R-12 can be reused, provided that it is cleaned to specifications in SAE J1991. The purpose of this document is to establish the specific minimum equipment specifications required for recycle of R-12 that has been directly removed from mobile systems for reuse in mobile automotive air-conditioning systems.

1. SCOPE:

The purpose of this document is to provide equipment specifications for CFC-12 (R-12) recycling and/or recovery, and recharging systems. This information applies to equipment used to service automobiles, light trucks, and other vehicles with similar CFC-12 systems. Systems used on mobile vehicles for refrigerated cargo that have hermetically sealed systems are not covered in this document.

2. REFERENCES:

SAE J51, Automotive Air-Conditioning Hose

SAE J1991, Standard of Purity for Use in Mobile Air-Conditioning Systems

UL 1963 Section 40 Tests Service Hoses for Refrigerant-12 (Underwriters Laboratories)

Pressure Relief Device Standard Part 1 - Cylinders for Compressed Gases, LGA Pamphlet S-1.1

3. SPECIFICATION AND GENERAL DESCRIPTION:

3.1 The equipment must be able to extract and process R-12 from mobile air-conditioning systems to purity levels specified in SAE J1991.

3.2 The equipment shall be suitable for use in an automotive service garage environment as defined in 7.8.

3.3 The equipment must be certified by Underwriters Laboratories or an equivalent certifying laboratory.

4. REFRIGERATION RECYCLE EQUIPMENT REQUIREMENTS:

4.1 Moisture and Acid: The equipment shall incorporate a desiccant package that must be replaced before saturated with moisture and whose acid capacity is at least 5% by weight of total system dry desiccant.

4.1.1 The equipment shall be provided with a moisture detection means that will reliably indicate when moisture in the R-12 exceeds the allowable level and requires the filter/dryer replacement.

4.2 Filter: The equipment shall incorporate an in-line filter that will trap particulates of 15 µm spherical diameter or greater.

4.3 Noncondensable Gas:

4.3.1 If the equipment has a self-contained recovery tank, a device is required to alert the operator that the noncondensable level has been exceeded.

4.3.2 Transfer of Recycled Refrigerant: Recycled refrigerant for recharging and transfer shall be taken from the liquid phase only.

5. SAFETY REQUIREMENTS:

5.1 The equipment must comply with applicable federal, state and local requirements on equipment related to the handling of R-12 material. Safety precautions or notices related to the safe operation of the equipment shall be prominently displayed on the equipment and should also state "Caution Should Be Operated By Qualified Personnel".

6. OPERATING INSTRUCTIONS:

6.1 The equipment manufacturer must provide operating instructions, necessary maintenance procedures, and source information for replacement parts and repair.

6.2 The equipment must prominently display the manufacturer's name, address and any items that require maintenance or replacement that affect the proper operation of the equipment. Operation manuals must cover information for complete maintenance of the equipment to assure proper operation.

7. FUNCTIONAL DESCRIPTION:

7.1 The equipment must be capable of ensuring recovery of the R-12 from the system being serviced, by reducing the system to a vacuum.

7.2 To prevent overcharge, the equipment must be equipped to protect the tank used to store the recycled refrigerant with a shutoff device and a mechanical pressure relief valve.

7.3 Portable refillable tanks or containers used in conjunction with this equipment must meet applicable Department of Transportation (DOT) or Underwriters Laboratories (UL) Standards and be adaptable to existing refrigerant service and charging equipment.

7.4 During the recovery and/or recycle, the equipment must provide overfill protection to assure that during filling or transfer, the tank or storage container cannot exceed 80% of volume at 70°F (21.1°C) of its maximum rating as defined by DOT standards, CFR Title 49 Part/Section 173.304 and American Society of Mechanical Engineers.

7.4.1 Additional Storage Tank Requirements:

7.4.1.1 The cylinder valve shall comply with the standard for cylinder valves, UL 1769.

7.4.1.2 The pressure relief device shall comply with the Pressure Relief Device Standard Part 1 - Cylinders for Compressed Gases, CGA Pamphlet S-1.1.

7.4.1.3 The tank assembly shall be marked to indicate the first retest date, which shall be 5 years after date of manufacture. The marking shall indicate that retest must be performed every subsequent 5 years. The marking shall be in letters at least 1/4 in high.

7.5 All flexible hoses must meet SAE J51 or UL 1963 Section 40.

7.6 Service hoses must have shutoff devices located within 12 in (30 cm) of the connection point to the system being serviced to minimize introduction of noncondensable gases into the recovery equipment and the release of the refrigerant when being disconnected.

7.7 The equipment must be able to separate the lubricant from recovered refrigerant and accurately indicate in 1 oz units (28 grams).

7.8 The equipment must be capable of continuous operation in ambient of 50 to 120°F (10 to 49°C).

7.9 The equipment must be compatible with leak detection material that may be present in the mobile AC system.

8. TESTING:

This test procedure and the requirements are used for evaluation of the equipment for its ability to clean the contaminated R-12 refrigerant.

8.1 The equipment shall clean the contaminated R-12 refrigerant to the minimum purity level as defined in SAE J1991, when tested in accordance with the following conditions:

8.2 For test validation, the equipment is to be operated according to the manufacturer's instructions.

8.3 The equipment must be preconditioned with 30 lb (13.6 kg) of the standard contaminated R-12 at an ambient of 70°F (21°C) before starting the test cycle. Sample amounts are not to exceed 2.5 lb (1.13 kg) with sample amounts to be repeated every 5 min. The sample method fixture, defined in Fig. 1, shall be operated at 75°F (24°C).

8.4 Contaminated R-12 Samples:

8.4.1 Standard contaminated R-12 refrigerant shall consist of liquid R-12 with 100 ppm (by weight) moisture at 70°F and 45 000 ppm (by weight) mineral oil 525 suspension viscosity nominal and 770 ppm by weight of noncondensable gases (air).

8.4.2 High moisture contaminated sample shall consist of R-12 vapor with 1000 ppm (by weight) moisture.

8.4.3 High oil contaminated sample shall consist of R-12 with 200 000 ppm (by weight) mineral oil 525 suspension viscosity nominal.

8.5 Test Cycle:

8.5.1 After preconditioning as stated in 8.3, the test cycle is started, processing the following contaminated samples through the equipment:

8.5.1.1 30 lb (13.6 kg) of standard contaminated R-12.

8.5.1.2 2.2 lb (1 kg) of high oil contaminated R-12.

8.5.1.3 10 lb (4.5 kg) of standard contaminated R-12.

8.5.1.4 2.2 lb (1 kg) of high moisture contaminated R-12.

8.6 Equipment Operating Ambient:

8.6.1 The R-12 is to be cleaned to the minimum purity level, as defined in SAE J1991, with the equipment operating in a stable ambient of 50, 70, and 120°F (10, 21, and 49°C) and processing the samples as defined in 8.5.

8.7 Sample Analysis:

8.7.1 The processed contaminated samples shall be analyzed according to the following procedure.

8.8 Quantitative Determination of Moisture:

8.8.1 The cleaned sample of R-12 is to be subjected to a quantitative determination of the moisture content by Karl Fischer titration.

8.8.2 The apparatus employed is a Karl Fischer coulometer, an automated instrument for precise determination of small amounts of water. The weighed sample of liquid R-12 is introduced directly into the analyte of the Karl Fischer coulometer. A coulometric titration by the instrument is conducted and the results are calculated and displayed as parts per million moisture weight.

8.9 Determination of Percent Oil:

8.9.1 The amount of oil in the cleaned sample of R-12 is to be determined by gravimetric analysis.

8.9.2 A weighed 100 mL sample of the liquid R-12 is placed in a preweighted graduated Goetz phosphorous tube of 100 mL nominal capacity. The sample and containing tube are maintained in ambient air at a minimum temperature of 90°F (32°C) above the expected boiling point of the refrigerant. When 85 mL of the sample has evaporated, the tube is then immersed in a refrigerated brine bath at a temperature of 50°F (10°C) above the boiling point of the sample for 30 min. The residual sample, if any, is allowed to reach room temperature. The tube is reweighed and the percent of oil is calculated.

8.10 Noncondensable Gas:

8.10.1 The sample is to be analyzed using gas chromatography to determine the noncondensable gas content. The cleaned refrigerant is to be sampled in the liquid phase through a closed loop or by an airtight syringe into the injector.

8.11 Sample Requirements:

8.11.1 The sample shall be tested as defined in 8.7, 8.8, 8.9, and 8.10 at ambient temperatures of 50, 70, and 120°F (10, 21, and 49°C) as defined in 8.6.1.

9. DATE OF EFFECTIVENESS:

This recommended practice will become a standard after one year.

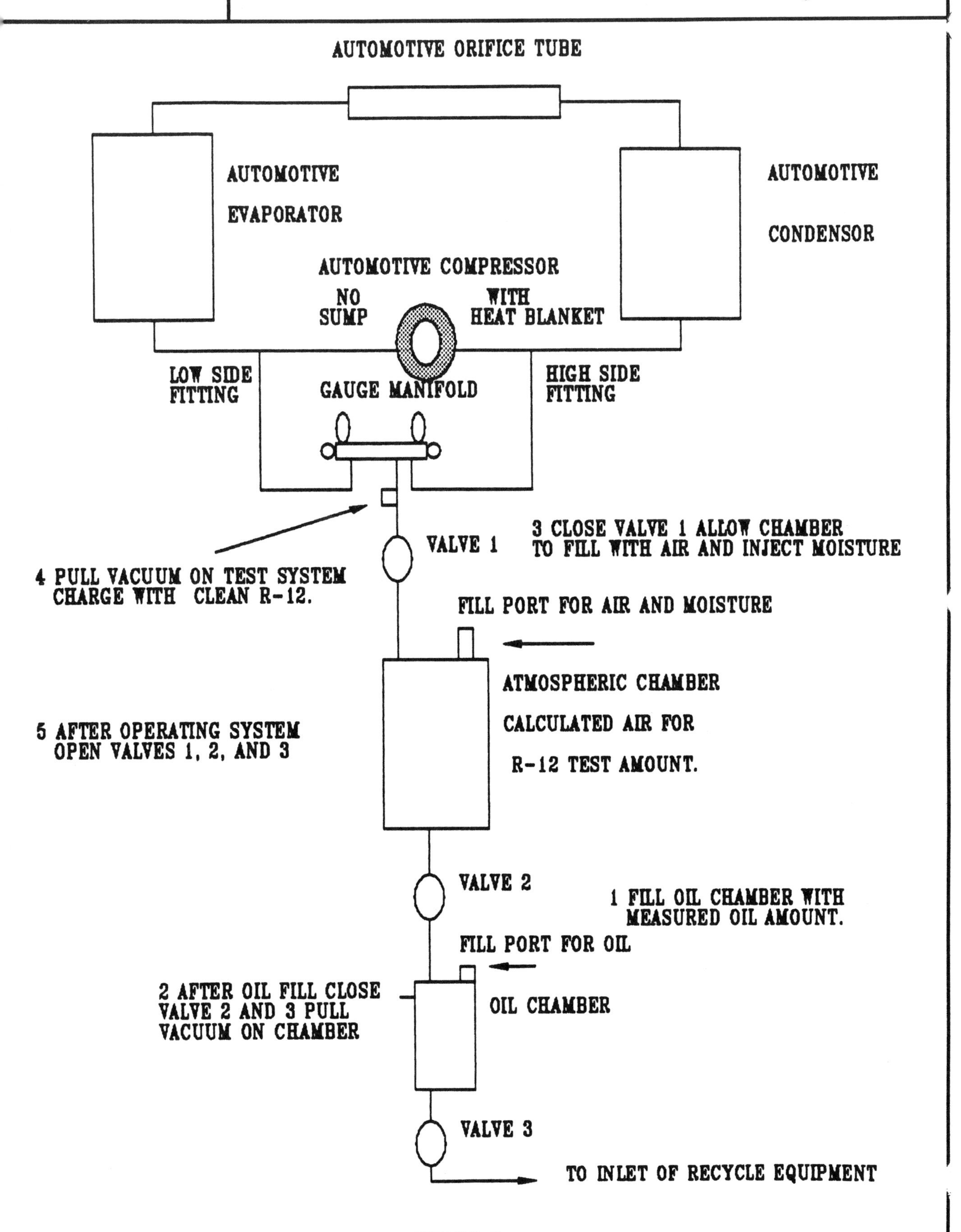
AUTOMOTIVE ORIFICE TUBE
AUTOMOTIVE
EVAPORATOR
AUTOMOTIVE
CONDENSOR
AUTOMOTIVE COMPRESSOR
NO
SUMP
WITH
HEAT BLANKET
LOW SIDE
FITTING
GAUGE MANIFOLD
HIGH SIDE
FITTING
VALVE 1
3 CLOSE VALVE 1 ALLOW CHAMBER
TO FILL WITH AIR AND INJECT MOISTURE
4 PULL VACUUM ON TEST SYSTEM
CHARGE WITH CLEAN R-12.
FILL PORT FOR AIR AND MOISTURE
ATMOSPHERIC CHAMBER
CALCULATED AIR FOR
R-12 TEST AMOUNT.
5 AFTER OPERATING SYSTEM
OPEN VALVES 1, 2, AND 3
VALVE 2
1 FILL OIL CHAMBER WITH
MEASURED OIL AMOUNT.
FILL PORT FOR OIL
2 AFTER OIL FILL CLOSE
VALVE 2 AND 3 PULL
VACUUM ON CHAMBER
OIL CHAMBER
VALVE 3
TO INLET OF RECYCLE EQUIPMENT

FIGURE 1

RATIONALE:

Not applicable.

RELATIONSHIP OF SAE STANDARD TO ISO STANDARD:

Not applicable.

REFERENCE SECTION:

SAE J51, Automotive Air-Conditioning Hose

SAE J1991, Standard of Purity for Use in Mobile Air-Conditioning Systems

UL 1963 Section 40 Tests Service Hoses for Refrigerant-12 (Underwriters Laboratories)

Pressure Relief Device Standard Part 1 - Cylinders for Compressed Gases, LGA Pamphlet S-1.1

APPLICATION:

The purpose of this document is to provide equipment specifications for CFC-12 (R-12) recycling and/or recovery, and recharging systems. This information applies to equipment used to service automobiles, light trucks and other vehicles with similar CFC-12 systems. Systems used on mobile vehicles for refrigerated cargo which have hermetically sealed systems are not covered in this document.

COMMITTEE COMPOSITION:

DEVELOPED BY THE SAE DEFROST AND INTERIOR CLIMATE CONTROL STANDARDS COMMITTEE:

W. J. Atkinson, Sun Test Engineering, Paradise Valley, AZ - Chairman
J. J. Amin, Union Lake, MI
H. S. Andersson, Saab Scania, Sweden
P. E. Anglin, ITT Higbie Mfg. Co., Rochester, MI
R. W. Bishop, GMC, Lockport, NY
D. Hawks, General Motors Corporation, Pontiac, MI
J. J. Hernandez, NAVISTAR, Ft. Wayne, IN
H. Kaltner, Volkswagen AG, Germany, Federal Republic
D. F. Last, GMC, Troy, MI
D. E. Linn, Volkswagen of America, Warren, MI
J. H. McCorkel, Freightliner Corp., Charlotte, NC
C. J. McLachlan, Livonia, MI
H. L. Miner, Climate Control Inc., Decatur, IL
R. J. Niemiec, General Motors Corp., Pontiac, MI
N. Novak, Chrysler Corp., Detroit, MI
S. Oulouhojian, Mobile Air Conditioning Society, Upper Darby, PA
J. Phillips, Air International, Australia
R. H. Proctor, Murray Corp., Cockeysville, MD
G. Rolling, Behr America Inc., Ft. Worth, TX
C. D. Sweet, Signet Systems Inc., Harrodsburg, KY
J. P. Telesz, General Motors Corp., Lockport, NY

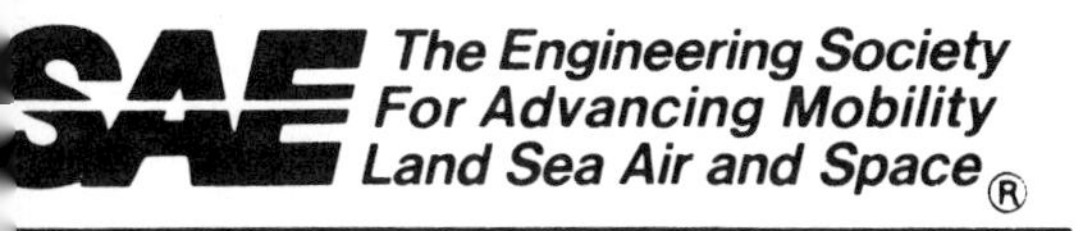
00 COMMONWEALTH DRIVE, WARRENDALE, PA 15096

HIGHWAY VEHICLE STANDARD

SAE J1991

Issued October 1989

Submitted for recognition as an American National Standard

STANDARD OF PURITY FOR USE IN MOBILE AIR-CONDITIONING SYSTEMS

FOREWORD: Due to the CFC's damaging effect on the ozone layer, recycle of CFC-12 (R-12) used in mobile air-conditioning systems is required to reduce system venting during normal service operations. Establishing recycle specifications for R-12 will assure that system operation with recycled R-12 will provide the same level of performance as new refrigerant.

Extensive field testing with the EPA and the auto industry indicate that reuse of R-12 removed from mobile air-conditioning systems can be considered, if the refrigerant is cleaned to a specific standard. The purpose of this standard is to establish the specific minimum levels of R-12 purity required for recycled R-12 removed from mobile automotive air-conditioning systems.

1. SCOPE:

This information applies to refrigerant used to service automobiles, light trucks, and other vehicles with similar CFC-12 systems. Systems used on mobile vehicles for refrigerated cargo that have hermetically sealed, rigid pipe are not covered in this document.

2. REFERENCES

SAE J1989, Recommended Service Procedure for the Containment of R-12

SAE J1990, Extraction and Recycle Equipment for Mobile Automotive Air-Conditioning Systems

ARI Standard 700-88

3. PURITY SPECIFICATION:

The refrigerant in this document shall have been directly removed from, and intended to be returned to, a mobile air-conditioning system. The contaminants in this recycled refrigerant 12 shall be limited to moisture, refrigerant oil, and noncondensable gases, which shall not exceed the following level:

3.1 Moisture: 15 ppm by weight.

3.2 Refrigerant Oil: 4000 ppm by weight.

3.3 Noncondensable Gases (Air): 330 ppm by weight.

4. REFRIGERATION RECYCLE EQUIPMENT USED IN DIRECT MOBILE AIR-CONDITIONING SERVICE OPERATIONS REQUIREMENT:

4.1 The equipment shall meet SAE J1990, which covers additional moisture, acid, and filter requirements.

4.2 The equipment shall have a label indicating that it is certified to meet this document.

5. PURITY SPECIFICATION OF RECYCLED R-12 REFRIGERANT SUPPLIED IN CONTAINERS FROM OTHER RECYCLE SOURCES:

Purity specification of recycled R-12 refrigerant supplied in containers from other recycle sources, for service of mobile air-conditioning systems, shall meet ARI Standard 700-88 (Air Conditioning and Refrigeration Institute).

6. OPERATION OF THE RECYCLE EQUIPMENT:

This shall be done in accordance with SAE J1989.

RATIONALE:

Not applicable.

RELATIONSHIP OF SAE STANDARD TO ISO STANDARD:

Not applicable.

REFERENCE SECTION:

SAE J1989, Recommended Service Procedure for the Containment of R-12

SAE J1990, Extraction and Recycle Equipment for Mobile Automotive Air-Conditioning Systems

ARI Standard 700-88

APPLICATION:

This information applies to refrigerant used to service automobiles, light trucks, and other vehicles with similar CFC-12 systems. Systems used on mobile vehicles for refrigerated cargo that have hermetically sealed, rigid pipe are not covered in this document.

COMMITTEE COMPOSITION:

DEVELOPED BY THE SAE DEFROST AND INTERIOR CLIMATE CONTROL STANDARDS COMMITTEE:

W. J. Atkinson, Sun Test Engineering, Paradise Valley, AZ - Chairman
J. J. Amin, Union Lake, MI
H. S. Andersson, Saab Scania, Sweden
P. E. Anglin, ITT Higbie Mfg. Co., Rochester, MI
R. W. Bishop, GMC, Lockport, NY
D. Hawks, General Motors Corporation, Pontiac, MI
J. J. Hernandez, NAVISTAR, Ft. Wayne, IN
H. Kaltner, Volkswagen AG, Germany, Federal Republic
D. F. Last, GMC, Troy, MI
D. E. Linn, Volkswagen of America, Warren, MI
J. H. McCorkel, Freightliner Corp., Charlotte, NC
C. J. McLachlan, Livonia, MI
H. L. Miner, Climate Control Inc., Decatur, IL
R. J. Niemiec, General Motors Corp., Pontiac, MI
N. Novak, Chrysler Corp., Detroit, MI
S. Oulouhojian, Mobile Air Conditioning Society, Upper Darby, PA
J. Phillips, Air International, Australia
R. H. Proctor, Murray Corp., Cockeysville, MD
G. Rolling, Behr America Inc., Ft. Worth, TX
C. D. Sweet, Signet Systems Inc., Harrodsburg, KY
J. P. Telesz, General Motors Corp., Lockport, NY

CHAPTER ELEVEN

QUESTIONS

While researching this book, I heard many of the same questions asked over and over. This chapter is dedicated to some of these questions, as well as the answers to these questions.

Isn't this entire ozone layer breakdown a marketing maneuver on the part of the refrigerant manufacturers?

NO!! It is a genuine problem that demands immediate concern.

Is the ozone layer continuing to break down?

A report released by the United Nations Environment Program shows that the ozone layer is breaking down faster than previously thought. In addition, the breakdown is not occurring just over the poles of the earth, it is occurring just about everywhere except for a tropical belt around the earth. Data released by NASA indicates that the presence of chlorine in the atmosphere is far worse than anyone imagined.

At what rate is the ozone layer breaking down?
Estimates indicate that over the last 12 years, a 3% breakdown has occurred, happening primarily in the summer. Some estimates believe this number to be closer to a 4-5% breakdown since 1978. On going studies show the numbers to be much worse.

Isn't the phaseout of CFCs going to be pushed forward?

DuPont has announced that they will phase out CFC production by the end of 1994. This is ahead of the planned phaseout under the Montreal Protocol. In addition, R-22 sales for all but service applications may be discontinued by January 1, 2005 in the developed world. There is a rising movement to accelerate the phaseout of R-22 even earlier. At the same time, there is a growing concern that supplies of HCFCs may not be sufficient to meet future service needs. However, acceptable alternative refrigerants are not yet available to make the changeover. Industry and the government are just beginning to study this concern and much remains to be done before any workable solution can be implemented.

What is going to happen to R-502?

There has been research on a direct replacement and also a blend that is composed of substitutes for R-502.

A salesman tried to sell me a machine that calls itself a reclaim rather than a recycle machine. Is that possible?

No. In order for a refrigerant to be considered reclaimed, it must be laboratory tested. Even if a portable machine could distill refrigerant to virgin standards, the operator still could not determine in the field if the refrigerant is pure enough to be considered virgin.

Are there other methods of solving the ozone crisis?

From scientist to cranks, many methods of reducing chlorine emissions have been proposed. One group has called for mountaintop lasers to "zap" the molecules as they ascend. Still another group has proposed spraying tons of chlorine scavenging hydrocarbons into the atmosphere to attack the chlorine. None are considered feasible at this point.

Is there testing going on for certification standards on hvac/r recycling equipment?

ARI is making arrangements with an independent laboratory to test equipment relative to ARI Standard 740-91.

Since the hvac/r industry is cleaning up its act will the tax on R-12 be repealed?

No, the tax will stand. It is an excellent revenue source. And, there might also be some additional taxes imposed on "greenhouse" gas emissions.

Are the states looking to regulate the hvac/r and automotive industries in addition to the regulations imposed by the federal government?

Many states are considering legislation that will affect hvac/r and automotive service. These rules would be in addition to the EPA rules. If these rules are tougher than the EPA standards, then the state rules will override the federal rule, with EPA approval. Trying to operate a refrigeration service may become very confusing in the future due to the possibility of many rules governing operation.

From a service standpoint, what can a technician do to help improve the ozone problem?

The technician can find and repair leaks in every system that requires service. With many of the new refrigerants, current leak detection will not work, but manufacturers are working on this. Possible methods for detecting new refrigerant leaks include using fluorescent dyes and developing compound-specific detectors to sense the new refrigerants.

What about the lubrication problems of the new refrigerants?

Refrigerant manufacturers are working to resolve the lubrication issue. R-123 will work with a mineral-based oil, modified PAG oils are the proposed solution for R-134a and lubrication alternatives are being discovered that will work with the blends. One manufacturer feels that the solution to the oil problem with R-134a does not lie in mineral or PAG oils, but rather in using an ester-based oil.

It should be understood that this entire area is new. Few units are actually on line using the alternatives. Research is occurring all over the globe. However, for an individual application, **consult with the manufacturer of the unit and follow the manufacturer's recommendation**. From a service standpoint, it is the warranty that is a prime concern on a changeout or retrofit. Everything we know now may change as experience begins to teach us all. The successful contractor in the new era of regulation will be the informed contractor.

Is R-134a going to be used in automotive air conditioning?

In 1992, several import manufacturers, including Mercedes-Benz, will offer R-134a. In the United States, Jeep Eagle has announced that a vehicle with R-134a will be forthcoming, and other manufacturers will follow suit very quickly. It is important to remember that R-134a and R-12 can never be mixed in equipment.

Does the CAA order purge units to be installed on chillers?

Chillers purge on a regular basis for non-condensables. There are units on the market that will capture the purged gases including the refrigerant that was released during the purge. These units should preclude a purge being considered an illegal emission. The CAA does not define every conceivable type of emission, but it is reasonable to assume that uncontrolled purging is illegal venting. ARI has been asked to develop a standard for purge equipment that is used with low-pressure chillers.

When certification is formerly required, will EPA certification be all that is necessary?

Many states, cities and counties are establishing their own certification programs. Technicians performing in certain areas may be required to obtain regional certification. More than likely, if the regional program is tougher than the EPA program, the regional program will supersede the EPA program.

What is SUVA Cold-Mp?

R-134a is sold by DuPont under the brand name SUVA Cold-Mp.

What is the difference between ARI 740-91 and ARI 700-88?

ARI 740-91 sets standards to certify recycling equipment, while ARI 700-88 establishes a laboratory requirement for refrigerant purity. ARI 700-88 does not apply to situations in which the technician is returning the filtered refrigerant to the unit (recycling). ARI 700-88 looks at the refrigerant from a laboratory standard of purity and is not concerned with the capability of the recycling machine unless the batch is analyzed in a laboratory. ARI 740-91 is more concerned with the recycling equipment itself. Appendix B and C offer copies of both documents.

APPENDIX A

Stratospheric Ozone Protection Final Rule Summary
Issued by the United States Environmental Protection Agency

United States
Environmental Protection
Agency

Air and Radiation
(6205J)

EPA-430-F-93-010
June 1993

EPA

Stratospheric Ozone Protection

Final Rule Summary

COMPLYING WITH THE REFRIGERANT RECYCLING RULE

This fact sheet provides an overview of the refrigerant recycling requirements of section 608 of the Clean Air Act, 1990, as amended (CAA), including final regulations published on May 14, 1993 (58 FR 28660), and the prohibition on venting that became effective on July 1, 1992.

TABLE OF CONTENTS

Overview

Under section 608 of the CAA, EPA has established regulations that:

- Require service practices that maximize recycling of ozone-depleting compounds (both chlorofluorocarbons [CFCs] and hydrochlorofluorocarbons [HCFCs]) during the servicing and disposal of air-conditioning and refrigeration equipment.
- Set certification requirements for recycling and recovery equipment, technicians, and reclaimers.
- Restrict the sale of refrigerant to certified technicians.
- Require persons servicing or disposing of air-conditioning and refrigeration equipment to certify to EPA that they have acquired recycling or recovery equipment and are complying with the requirements of the rule.
- Require the repair of substantial leaks in air-conditioning and refrigeration equipment with a charge of greater than 50 pounds.
- Establish safe disposal requirements to ensure removal of refrigerants from

goods that enter the waste stream with the charge intact (e.g., motor vehicle air conditioners, home refrigerators, and room air conditioners).

The Prohibition on Venting

Effective July 1, 1992, section 608 of the Act prohibits individuals from knowingly venting ozone-depleting compounds used as refrigerants into the atmosphere while maintaining, servicing, repairing, or disposing of air-conditioning or refrigeration equipment. Only four types of releases are permitted under the prohibition:

1. "De minimis" quantities of refrigerant released in the course of making good faith attempts to recapture and recycle or safely dispose of refrigerant.

2. Refrigerants emitted in the course of normal operation of air-conditioning and refrigeration equipment (as opposed to during the maintenance, servicing, repair, or disposal of this equipment) such as from mechanical purging and leaks. However, EPA is requiring the repair of substantial leaks.

3. Mixtures of nitrogen and R-22 that are used as holding charges or as leak test gases, because in these cases, the ozone-depleting compound is not used as a refrigerant. However, a technician may not avoid recovering refrigerant by adding nitrogen to a charged system; before nitrogen is added, the system must be evacuated to the appropriate level in Table 1. Otherwise, the CFC or HCFC vented along with the nitrogen will be considered a refrigerant. Similarly, *pure* CFCs or HCFCs released from appliances will be presumed to be refrigerants, and their release will be considered a violation of the prohibition on venting.

4. Small releases of refrigerant which result from purging hoses or from connecting or disconnecting hoses to charge or service appliances will not be considered violations of the prohibition on venting. However, recovery and recycling equipment manufactured after November 15, 1993, must equipped with low-loss fittings.

Regulatory Requirements

Service Practice Requirements

1. Evacuation Requirements. Beginning July 13, 1993, technicians are required to evacuate air-conditioning and refrigeration equipment to established vacuum levels. If the technician's recovery or recycling equipment is manufactured any time before November 15, 1993, the air-conditioning and refrigeration equipment must be evacuated to the levels described in the first column of Table 1. If the technician's recovery or recycling equipment is manufactured on or after November 15, 1993, the air-conditioning and refrigeration equipment must be evacuated to the levels described in the second column of Table 1, and the recovery or recycling equipment must have been certified by an EPA-approved equipment testing organization (see *Equipment Certification*, below).

Technicians repairing small appliances, such as household refrigerators, household freezers, and water coolers, are required to recover 80-90 percent of the refrigerant in the system, depending on the status of the system's compressor.

TABLE 1
REQUIRED LEVELS OF EVACUATION FOR APPLIANCES
EXCEPT FOR SMALL APPLIANCES, MVACS, AND MVAC-LIKE APPLIANCES

Type of Appliance	Inches of Mercury Vacuum* Using Equipment Manufactured:	
	Before Nov. 15, 1993	On or after Nov. 15, 1993
HCFC-22 appliance** normally containing less than 200 pounds of refrigerant	0	0
HCFC-22 appliance** normally containing 200 pounds or more of refrigerant	4	10
Other high-pressure appliance** normally containing less than 200 pounds of refrigerant (CFC-12, -500, -502, -114)	4	10
Other high-pressure appliance** normally containing 200 pounds or more of refrigerant (CFC-12, -500, -502, -114)	4	15
Very High Pressure Appliance (CFC-13, -503)	0	0
Low-Pressure Appliance (CFC-11, HCFC-123)	25	25 mm Hg absolute

*Relative to standard atmospheric pressure of 29.9" Hg.
**Or isolated component of such an appliance

2. *Exceptions to Evacuation Requirements.* EPA has established limited exceptions to its evacuation requirements for 1) repairs to leaky equipment and 2) repairs that are not major and that are not followed by an evacuation of the equipment to the environment.

If, due to leaks, evacuation to the levels in Table 1 is not attainable, or would substantially contaminate the refrigerant being recovered, persons opening the appliance must:

- isolate leaking from non-leaking components wherever possible;
- evacuate non-leaking components to the levels in Table 1; and
- evacuate leaking components to the lowest level that can be attained without substantially contaminating the refrigerant. This level cannot exceed 0 psig.

If evacuation of the equipment to the environment is not to be performed when repairs are complete, and if the repair is not major, then the appliance must:

- be evacuated to at least 0 psig before it is opened if it is a high- or very high-pressure appliance; or

- be pressurized to 0 psig before it is opened if it is a low-pressure appliance. Methods that require subsequent purging (e.g., nitrogen) cannot be used.

"Major" repairs are those involving removal of the compressor, condenser, evaporator, or auxiliary heat exchanger coil.

3. Reclamation Requirement. EPA has also established that refrigerant recovered and/or recycled can be returned to the same system or other systems owned by the same person without restriction. If refrigerant changes ownership, however, that refrigerant must be reclaimed (i.e., cleaned to the ARI 700 standard of purity and chemically analyzed to verify that it meets this standard). This provision will expire in May, 1995, when it may be replaced an off-site recycling standard.

Equipment Certification

The Agency has established a certification program for recovery and recycling equipment. Under the program, EPA requires that equipment manufactured on or after November 15, 1993, be tested by an EPA-approved testing organization to ensure that it meets EPA requirements. Recycling and recovery equipment intended for use with air-conditioning and refrigeration equipment besides small appliances must be tested under the ARI 740-1993 test protocol, which is included in the final rule as Appendix B. Recovery equipment intended for use with small appliances must be tested under either the ARI 740-1993 protocol or Appendix C of the final rule. The Agency is requiring recovery efficiency standards that vary depending on the size and type of air-conditioning or refrigeration equipment being serviced. For recovery and recycling equipment intended for use with air-conditioning and refrigeration equipment besides small appliances, these standards are the same as those in the second column of Table 1. Recovery equipment intended for use with small appliances must be able to recover 90 percent of the refrigerant in the small appliance when the small appliance compressor is operating and 80 percent of the refrigerant in the small appliance when the compressor is not operating.

Equipment Grandfathering

Equipment manufactured before November 15, 1993, including home-made equipment, will be grandfathered if it meets the standards in the first column of Table 1. Third-party testing is not required for equipment manufactured before November 15, 1993, but equipment manufactured on or after that date, including home-made equipment, must be tested by a third-party (see *Equipment Certification* above).

Refrigerant Leaks

Owners of equipment with charges of greater than 50 pounds are required to repair substantial leaks. A 35 percent annual leak rate is established for the industrial process and commercial refrigeration sectors as the trigger for requiring repairs. An annual leak rate of 15 percent of charge per year is established for comfort cooling chillers and all other equipment with a charge of over 50 pounds other than industrial process and commercial refrigeration equipment. Owners of air-conditioning and refrigeration equipment with more than 50 pounds of charge must keep records of the quantity of refrigerant added to their equipment during servicing and maintenance procedures.

Mandatory Technician Certification

EPA has established a mandatory technician certification program. The Agency has developed four types of certification:

- For servicing small appliances (Type I).
- For servicing or disposing of high- or very high-pressure appliances, except small appliances and MVACs (Type II).
- For servicing or disposing of low-pressure appliances (Type III)
- For servicing all types of equipment (Universal).

Persons removing refrigerant from small appliances and motor vehicle air conditioners for purposes of **disposal** of these appliances do not have to be certified.

Technicians are required to pass an EPA-approved test given by an EPA-approved certifying organization to become certified under the mandatory program. Technicians must be certified by November 14, 1994. EPA expects to have approved some certifying organizations by September of this year. The Stratospheric Ozone Hotline will distribute lists of approved organizations at that time.

EPA plans to "grandfather" individuals who have already participated in training and testing programs provided the testing programs 1) are approved by EPA and 2) provide additional, EPA-approved materials or testing to these individuals to ensure that they have the required level of knowledge.

Although any organization may apply to become an approved certifier, EPA plans to give priority to national organizations able to reach large numbers of people. EPA encourages smaller training organizations to make arrangements with national testing organizations to administer certification examinations at the conclusion of their courses.

Refrigerant Sales Restrictions

Under Section 609 of the Clean Air Act, sales of CFC-12 in containers smaller than 20 pounds are now restricted to technicians certified under EPA's motor vehicle air conditioning regulations. Persons servicing appliances other than motor vehicle air conditioners may still buy containers of CFC-12 larger than 20 pounds.

After November 14, 1994, the sale of refrigerant in any size container will be restricted to technicians certified either under the program described in *Technician Certification* above or under EPA's motor vehicle air conditioning regulations.

Certification by Owners of Recycling and Recovery Equipment

EPA is requiring that persons servicing or disposing of air-conditioning and refrigeration equipment certify to EPA that they have acquired (built, bought, or leased) recovery or recycling equipment and that they are complying with the applicable requirements of this rule. This certification must be signed by the owner of the equipment or another responsible officer and sent to the appropriate EPA Regional Office by August 12, 1993. A sample form for this certification is attached. Although owners of recycling and recovery equipment are required to list the number of trucks based at their shops, they do not need to have a piece of recycling or recovery equipment for every truck.

Reclaimer Certification

Reclaimers are required to return refrigerant to the purity level specified in ARI Standard 700-1988 (an industry-set purity standard) and to verify this purity using the laboratory protocol set forth in the same standard. In addition, reclaimers must release no more than 1.5 percent of the refrigerant during the reclamation process and must dispose of wastes properly. Reclaimers must certify by August 12, 1993, to the Section 608 Recycling Program Manager at EPA headquarters that they are complying with these requirements and that the information given is true and correct. The certification must also include the name and address of the reclaimer and a list of equipment used to reprocess and to analyze the refrigerant.

EPA encourages reclaimers to participate in third-party reclaimer certification programs, such as that operated by the Air-Conditioning and Refrigeration Institute (ARI). Third-party certification can enhance the attractiveness of a reclaimer's product by providing an objective assessment of its purity.

MVAC-like Appliances

Some of the air conditioners that are covered by this rule are identical to motor vehicle air conditioners (MVACs), but they are not covered by the MVAC refrigerant recycling rule (40 CFR Part 82 Subpart B) because they are used in vehicles that are not defined as "motor vehicles." These air conditioners include many systems used in construction equipment, farm vehicles, boats, and airplanes. Like MVACs in cars and trucks, these air conditioners typically contain two or three pounds of CFC-12 and use open-drive compressors to cool the passenger compartments of vehicles. (Vehicle air conditioners utilizing HCFC-22 are not included in this group and are therefore subject to the requirements outlined above for HCFC-22 equipment.) EPA is defining these air conditioners as "MVAC-like appliances" and is applying the MVAC rule's requirements for the certification and use of recycling and recovery equipment to them. That is, technicians servicing MVAC-like appliances must "properly use" recycling or recovery equipment that has been certified to meet the standards in Appendix A to 40 CFR Part 82, Subpart B. In addition, EPA is allowing technicians who service MVAC-like appliances to be certified by a certification program approved under the MVAC rule, if they wish.

Safe Disposal Requirements

Under EPA's rule, equipment that is typically dismantled on-site before disposal (e.g., retail food refrigeration, cold storage warehouse refrigeration, chillers, and industrial process refrigeration) has to have the refrigerant recovered in accordance with EPA's requirements for servicing. However, equipment that typically enters the waste stream with the charge intact (e.g., motor vehicle air conditioners, household refrigerators and freezers, and room air conditioners) is subject to special safe disposal requirements.

Under these requirements, the final person in the disposal chain (e.g., a scrap metal recycler or landfill owner) is responsible for ensuring that refrigerant is recovered from equipment before the final disposal of

OMB # 2060-0256
Expiration Date: 5/96

THE UNITED STATES ENVIRONMENTAL PROTECTION AGENCY (EPA) REFRIGERANT RECOVERY OR RECYCLING DEVICE ACQUISITION CERTIFICATION FORM

EPA regulations require establishments that service or dispose of refrigeration or air conditioning equipment to certify by August 12, 1993 that they have acquired recovery or recycling devices that meet EPA standards for such devices. To certify that you have acquired equipment, please complete this form according to the instructions and **mail it to the appropriate EPA Regional Office. BOTH THE INSTRUCTIONS AND MAILING ADDRESSES CAN BE FOUND ON THE REVERSE SIDE OF THIS FORM.**

PART 1: ESTABLISHMENT INFORMATION

Name of Establishment

Street

(Area Code) Telephone Number

City State Zip Code

Number of Service Vehicles Based at Establishment

County

PART 2: REGULATORY CLASSIFICATION

Identify the type of work performed by the establishment. **Check all boxes that apply.**

- ☐ Type A -Service small appliances
- ☐ Type B -Service refrigeration or air conditioning equipment other than small appliances
- ☐ Type C -Dispose of small appliances
- ☐ Type D -Dispose of refrigeration or air conditioning equipment other than small appliances

PART 3: DEVICE IDENTIFICATION

Name of Device(s) Manufacturer	Model Number	Year	Serial Number (if any)	Check Box if Self-Contained
1.				☐
2.				☐
3.				☐
4.				☐
5.				☐
6.				☐
7.				☐

PART 4: CERTIFICATION SIGNATURE

I certify that the establishment in Part 1 has acquired the refrigerant recovery or recycling device(s) listed in Part 2, that the establishment is complying with Section 608 regulations, and that the information given is true and correct.

Signature of Owner/Responsible Officer Date Name (Please Print) Title

Public reporting burden for this collection of information is estimated to vary from 20 minutes to 60 minutes per response with an average of 40 minutes per response, including time for reviewing instructions, searching existing data sources, gathering and maintaining the data needed, and completing the collection of information. Send comments regarding ONLY the burden estimates or any other aspects of this collection of information, including suggestions for reducing this burden to Chief, Information Policy Branch; EPA;401 M St., S.W. (PM-223Y); Washington, DC 20460; and to the Office of Information and Regulatory Affairs, Office of Management and Budget, Washington, DC 20503, marked "Attention: Desk Officer of EPA" DO NOT SEND THIS FORM TO THE ABOVE ADDRESSES. ONLY SEND COMMENTS TO THESE ADDRESSES.

Instructions

Part 1: Please provide the name, address, and telephone number of the establishment where the refrigerant recovery or recycling device(s) is (are) located. Please complete one form for each location. State the number of vehicles based at this location that are used to transport technicians and equipment to and from service sites.

Part 2: Check the appropriate boxes for the type of work performed by technicians who are employees of the establishment. The term "small appliance" refers to any of the following products that are fully manufactured, charged, and hermetically sealed in a factory with five pounds or less of refrigerant: refrigerators and freezers designed for home use, room air conditioners (including window air conditioners and packaged terminal air conditioners), packaged terminal heat pumps, dehumidifiers, under-the-counter ice makers, vending machines, and drinking water coolers.

Part 3: For each recovery or recycling device acquired, please list the name of the manufacturer of the device, and (if applicable) its model number and serial number.

If more than 7 devices have been acquired, please fill out an additional form and attach it to this one. Recovery devices that are self-contained should be listed first and should be identified by checking the box in the last column on the right. Self-contained recovery equipment means refrigerant recovery or recycling equipment that is capable of removing the refrigerant from an appliance without the assistance of components contained in the appliance. On the other hand, system-dependent recovery equipment means refrigerant recovery equipment that requires the assistance of components contained in an appliance to remove the refrigerant from the appliance.

If the establishment has been listed as Type B and/or Type D in Part 2, then the first device listed in Part 3 must be a self-contained device and identified as such by checking the box in the last column on the right.

If any of the devices are homemade, they should be identified by writing "homemade" in the column provided for listing the name of the device manufacturer. Type A or Type B establishments can use homemade devices manufactured before November 15, 1993. Type C or Type D establishments can use homemade devices manufactured anytime. If, however, a Type C or Type D establishment is using homemade equipment manufactured after November 15, 1993, then it must not use these devices for service jobs.

Part 4: This form must be signed by either the owner of the establishment or another responsible officer. The person who signs is certifying that the establishment has acquired the equipment, that the establishment is complying with Section 608 regulations, and that the information provided is true and correct.

EPA Regional Offices

Send your form to the EPA office listed under the state or territory in which the establishment is located.

Connecticut, Maine, Massachusetts, New Hampshire, Rhode Island, Vermont

> CAA 608 Enforcement Contact: EPA Region I, Mail Code APC, JFK Federal Building, One Congress Street, Boston, MA 02203

New York, New Jersey, Puerto Rico, Virgin Islands

> CAA 608 Enforcement Contact: EPA Region II, Jacob K. Javits Federal Building, Room 5000, 26 Federal Plaza, New York, NY 10278

Delaware, District of Columbia, Maryland, Pennsylvania, Virginia, West Virginia

> CAA 608 Enforcement Contact: EPA Region III, Mail Code 3AT21, 841 Chestnut Building, Philadelphia, PA 19107

Alabama, Florida, Georgia, Kentucky, Mississippi, North Carolina, South Carolina, Tennessee

> CAA 608 Enforcement Contact: EPA Region IV, Mail Code APT-AE, 345 Courtland Street, NE, Atlanta, GA 30365

Illinois, Indiana, Michigan, Minnesota, Ohio, Wisconsin

> CAA 608 Enforcement Contact: EPA Region V, Mail Code AT18J, 77 W. Jackson Blvd., Chicago, IL 60604

Arkansas, Louisiana, New Mexico, Oklahoma, Texas

> CAA 608 Enforcement Contact: EPA Region VI, Mail Code 6T-EC, First Interstate Tower at Fountain Place, 1445 Ross Ave., Suite 1200, Dallas TX 75202

Iowa, Kansas, Missouri, Nebraska

> CAA 608 Enforcement Contact: EPA Region VII, Mail Code ARTX/ARBR, 726 Minnesota Ave., Kansas City, KS 66101

Colorado, Montana, North Dakota, South Dakota, Utah, Wyoming

> CAA 608 Enforcement Contact: EPA Region VIII, Mail Code 8AT-AP, 999 18th Street, Suite 500, Denver, CO 80202

American Samoa, Arizona, California, Guam, Hawaii, Nevada

> CAA 608 Enforcement Contact: EPA Region IX, Mail Code A-3, 75 Hawthorne Street, San Francisco, CA 94105

Alaska, Idaho, Oregon, Washington

> CAA 608 Enforcement Contact: EPA Region X, Mail Code AT-082, 1200 Sixth Ave., Seattle, WA 98101

the equipment. However, persons "up-stream" can remove the refrigerant and provide documentation of its removal to the final person if this is more cost-effective.

The equipment used to recover refrigerant from appliances prior to their final disposal must meet the same "performance standards" as equipment used prior to servicing, but it does not need to be tested by a laboratory. This means that self-built equipment is allowed as long as it meets the performance requirements. For MVACs and MVAC-like appliances, the performance requirement is 102 mm of mercury vacuum and for small appliances, the recover equipment performance requirements are 90 percent efficiency when the appliance compressor is operational, and 80 percent efficiency when the appliance compressor is not operational.

Technician certification is not required for individuals removing refrigerant from appliances in the waste stream.

The safe disposal requirements are effective on July 13, 1993. The equipment must be registered or certified with the Agency by August 12, 1993. A sample form is attached.

Major Recordkeeping Requirements

Technicians servicing appliances that contain 50 or more pounds of refrigerant must provide the owner with an invoice that indicates the amount of refrigerant added to the appliance. Technicians must also keep a copy of their proof of certification at their place of business.

Owners of appliances that contain 50 or more pounds of refrigerant must keep servicing records documenting the date and type of service, as well as the quantity of refrigerant added.

Wholesalers who sell CFC and HCFC refrigerants must retain invoices that indicate the name of the purchaser, the date of sale, and the quantity of refrigerant purchased.

Reclaimers must maintain records of the names and addresses of persons sending them material for reclamation and the quantity of material sent to them for reclamation. This information must be maintained on a transactional basis. Within 30 days of the end of the calendar year, reclaimers must report to EPA the total quantity of material sent to them that year for reclamation, the mass of refrigerant reclaimed that year, and the mass of waste products generated that year.

Hazardous Waste Disposal

If refrigerants are recycled or reclaimed, they are not considered hazardous under federal law. In addition, used oils contaminated with CFCs are not hazardous on the condition that:

- They are not mixed with other waste.
- They are subjected to CFC recycling or reclamation.
- They are not mixed with used oils from other sources.

Used oils that contain CFCs after the CFC reclamation procedure, however, are subject to specification limits for used oil fuels if these oils are destined for burning. Individuals with questions regarding the proper handling of these materials should contact EPA's RCRA Hotline at 800-424-9346 or 703-920-9810.

Enforcement

EPA is performing random inspections, responding to tips, and pursuing potential cases against violators. Under the Act, EPA is authorized to assess fines of up to $25,000 per day for any violation of these regulations.

Planning and Acting for the Future

Observing the refrigerant recycling regulations for section 608 is essential in order to conserve existing stocks of refrigerants, as well as to comply with Clean Air Act requirements. However, owners of equipment that contains CFC refrigerants should look beyond the immediate need to maintain existing equipment in working order. **EPA urges equipment owners to act now and prepare for the phaseout of CFCs, which will be completed by January 1, 1996.** Owners are advised to begin the process of converting or replacing existing equipment with equipment that uses alternative refrigerants.

To assist owners, suppliers, technicians and others involved in comfort chiller and commercial refrigeration management, EPA has published a series of short fact sheets and expects to produce additional material. Copies of material produced by the EPA Stratospheric Protection Division are available from the Stratospheric Ozone Information Hotline (see hotline number below).

For Further Information

For further information concerning regulations related to stratospheric ozone protection, please call the Stratospheric Ozone Information Hotline: 800-296-1996. The Hotline is open between 10:00 AM and 4:00 PM, Eastern Time.

DEFINITIONS

Appliance
: Any device which contains and uses a class I (CFC) or class II (HCFC) substance as a refrigerant and which is used for household or commercial purposes, including any air conditioner, refrigerator, chiller, or freezer. EPA interprets this definition to include all air-conditioning and refrigeration equipment except that designed and used exclusively for military purposes.

Major maintenance, service, or repair
: Maintenance, service, or repair that involves removal of the appliance compressor, condenser, evaporator, or auxiliary heat exchanger coil.

MVAC-like appliance
: Mechanical vapor compression, open-drive compressor appliances used to cool the driver's or passenger's compartment of a non-road vehicle, including agricultural and construction vehicles. This definition excludes appliances using HCFC-22.

Reclaim
: To reprocess refrigerant to at least the purity specified in the ARI Standard 700-1988, Specifications for Fluorocarbon Refrigerants, and to verify this purity using the analytical methodology prescribed in the Standard.

Recover
: To remove refrigerant in any condition from an appliance and store it in an external container without necessarily testing or processing it in any way.

Recycle
: To extract refrigerant from an appliance and clean refrigerant for reuse without meeting all of the requirements for reclamation. In general, recycled refrigerant is refrigerant that is cleaned using oil separation and single or multiple passes through devices, such as replaceable core filter-driers, which reduce moisture, acidity, and particulate matter.

Self-contained recovery equipment
: Recovery or recycling equipment that is capable of removing the refrigerant from an appliance without the assistance of components contained in the appliance.

Small appliance
: Any of the following products that are fully manufactured, charged, and hermetically sealed in a factory with five pounds or less of refrigerant: refrigerators and freezers designed for home use, room air conditioners (including window air conditioners and packaged terminal air conditioners), packaged terminal heat pumps, dehumidifiers, under-the-counter ice makers, vending machines, and drinking water coolers.

System-dependent recovery equipment
: Recovery equipment that requires the assistance of components contained in an appliance to remove the refrigerant from the appliance.

Technician
: Any person who performs maintenance, service, or repair that could reasonably be expected to release class I (CFC) or class II (HCFC) substances into the atmosphere, including but not limited to installers, contractor employees, in-house service personnel, and in some cases, owners. Technician also means any person disposing of appliances except for small appliances.

TABLE 2
MAJOR RECYCLING RULE COMPLIANCE DATES

Requirement	Date
• Date after which owners of equipment containing more than 50 pounds of refrigerant with substantial leaks must have such leaks repaired.	June 14, 1993
• Evacuation requirements go into effect. • Recovery and recycling equipment requirements go into effect.	July 13, 1993
• Owners of recycling and recovery equipment must have certified to EPA that they have acquired such equipment and that they are complying with the rule. • Reclamation requirement goes into effect.	August 12, 1993
• All newly manufactured recycling and recovery equipment must be certified by an EPA-approved testing organization to meet the requirements in the second column of Table 1.	November 15, 1993
• All technicians must be certified. • Sales restriction goes into effect.	November 14, 1994
• Reclamation requirement expires.	May 14, 1995

United States
Environmental Protection
Agency
(6205-J)
Washington, DC 20460

APPENDIX B

1988 STANDARD for

SPECIFICATIONS FOR FLUOROCARBON REFRIGERANTS

Standard 700

1501 WILSON BOULEVARD • ARLINGTON, VIRGINIA 22209

IMPORTANT

Safety Recommendations

Procedures in this standard may involve hazardous materials, operations and equipment. This standard does not purport to address all the safety problems that might be associated with its use. It is the responsibility of whomever uses this standard to identify and establish appropriate safety and health practices and determine the applicability of regulatory limitations prior to use. Information may be obtained from ANSI/ASHRAE Standard 15, "Safety Code for Mechanical Refrigeration," the refrigerant manufacturers and other sources.

FOREWORD

The intent of this standard is to define a level of quality for new, reclaimed and/or repackaged refrigerants for use in new and existing refrigeration and air-conditioning products within the scope of ARI.

Contaminant limits were chosen to be within the sensitivity of recommended test methods, to be economically achievable by current processes and to provide satisfactory performance of these products.

This standard does not apply where refrigerant captured from a particular system is returned on site to the same system.

Price $12.50
Printed in U.S.A.

TABLE OF CONTENTS

SPECIFICATION FOR FLUOROCARBON REFRIGERANTS

Section 1. Purpose

1.1 *Purpose.* The purpose of this standard is to enable users to evaluate and accept/reject refrigerants regardless of source (new, reclaimed and/or repackaged) for use in new and existing refrigerating and air conditioning products within the scope of ARI.

1.1.1 This standard is intended for the guidance of the industry, including manufacturers, refrigerant reclaimers, repackagers, distributors, installers, servicemen, contractors and for consumers.

1.2 *Review and Amendment.* This standard is subject to review and amendment as the technology advances.

Section 2. Scope

2.1 *Scope.* This standard defines and classifies refrigerant contaminants primarily based on standard and generally available test methods and specifies acceptable levels of contaminants (purity requirements) for various fluorocarbon refrigerants regardless of source. These refrigerants are: R11; R12; R13; R22; R113; R114; R500; R502 and R503 as referenced in the ANSI/ASHRAE Standard "Number Designation of Refrigerants" (American Society of Heating, Refrigerating and Air Conditioning Engineers, Inc., Standard 34-78).

Section 3. Definitions

3.1 *"Shall", "Should", "Recommended", or "It Is Recommended".* "Shall", "should", "recommended", or "it is recommended" shall be interpreted as follows:

3.1.1 *Shall.* Where "shall" or "shall not" is used for a provision specified, that provision is mandatory if compliance with the standard is claimed.

3.1.2 *Should, Recommended, or It is Recommended.* "Should", "recommended", or "it is recommended" is used to indicate provisions which are not mandatory but which are desirable as good practice.

Section 4. Characterization of Refrigerants and Contaminants

4.1 Characterization of refrigerants and contaminants addressed are listed in the following general classifications:

4.1.1 *Characterization*
a. Boiling point
b. Boiling point range

4.1.2 *Contaminants*
a. Water
b. Chloride ion
c. Acidity
d. High boiling residue
e. Particulates/solids
f. Non-condensables
g. Other refrigerants

Section 5. Sampling, Test Methods and Maximum Permissible Contaminant Levels

5.1 The recommended referee test methods for the various contaminants are given in the following paragraphs. If alternate test methods are employed, the user must be able to demonstrate that they produce results equivalent to the specified referee method.

5.2 *Refrigerant Sampling.*

5.2.1 Special precautions should be taken to assure that representative samples are obtained for analysis. Sampling shall be done by trained laboratory personnel following accepted sampling and safety procedures.

5.2.2 *Gas Phase Sample.* A gas phase sample shall be obtained for determining the non-condensables by connecting the sample cylinder to an evacuated gas sampling bulb by means of a manifold. The manifold should have a valve arrangement that facilitates evacuation of all connecting tubing leading to the sampling bulb. Since non-condensable gases, if present, will concentrate in the vapor phase of the refrigerant, care must be exercised to eliminate introduction of air during the sample transfer. Purging is not an acceptable procedure for a gas phase sample since it may introduce a foreign product. Since R11 and R113 have normal boiling points at or above room temperature, non-condensable determination is not required for these refrigerants.

5.2.3 *Liquid Phase Sample.* A liquid phase sample, which may be obtained as follows, is required for all tests listed in this standard, except the test for non-condensables. Place an empty sample cylinder with the valve opened in an oven at 230°F [110°C] for one hour. Remove it from the oven while hot, immediately connect to an evacuation system and evacuate to less than 1 mm. mercury (1000 microns). Close the valve and allow it to cool.

5.2.3.1 The valve and lines from the unit to be sampled shall be clean and dry. Connect the line to the sample cylinder loosely. Purge through the loose connection. Make the connection tight at the end of the purge period. Take the sample as a liquid by chilling the sample cylinder slightly. Do not load the cylinder over 80 percent full at room temperature. This can be accomplished by weighing the empty cylinder and then the cylinder with refrigerant. The cylinder must not become completely full of liquid below 130°F [54.4°C]. When the desired amount of refrigerant has been collected, close the valve(s) and disconnect the sample cylinder immediately.

5.2.3.2 Check the sample cylinder for leaks and record the gross weight.

5.3 *Refrigerant Boiling Point and Boiling Range*

5.3.1 The test method shall be that described in the Federal Specification for "Fluorocarbon Refrigerants" BB-F-1421 B dated March 5, 1982, section 4.4.3.

5.3.2 The required values for boiling point and boiling point range are given in Table 1, "Physical Properties of Fluorocarbon Refrigerants and Maximum Contaminant Levels."

5.3.3 Gas chromatography (GC) is an acceptable alternate test method which can be used to characterize refrigerants. This is done by comparison to be known standards. Listed below are some readily available GC methods.

Alternate Gas Chromatography Test Methods

(See Appendix A for titles and sources)

Refrigerant	ICI	Dupont	Allied
R11	RSV/ALAB/CM3 and RSV/ALAB/CM4	F3205.165.01CW	G-11-7A
R12	RSV/ALAB/CM5	F3227.165.01CW(P)	G-12-7A
R13	RSV/ALAB/CM20	F3275.165.01CC(P)	—
R22	RSV/ALAB/CM8	F3290.165.01LV(P)	G-22-7A
R113	RSV/ALAB/CM6	F3297.165.01CC	GSVD-1A
R114	RSV/ALAB/CM21	F3305.165.01CC(P)	G-114-7A
R500	RSV/ALAB/CM5	F3327.165.01CW(P)	G-500-7A
R502	RSV/ALAB/CM8	F3333.165.01CC	G-502-7A
R503	RSV/ALAB/CM20	F3337.165.01CW(P)	G-503-7A

Note: Equivalent laboratory test methods may be available from other producers of these refrigerants.

5.4 *Water Content*

5.4.1 The Karl Fischer Test Method shall be used for determining the water content of refrigerant. This method is described in ASTM Standard for "Water In Gases Using Karl Fisher Reagent" E700-79, reapproved 1984 (American Society for Testing Materials, Philadelphia, PA). This method can be used for refrigerants that are either a liquid or a gas at room temperature, including Refrigerants 11 and 113. For all refrigerants, the sample for water analysis shall be taken from the liquid phase of the container to be tested. Proper operation of the analytical method requires special equipment and an experienced operator. The precision of the results is excellent if proper sampling and handling procedures are followed. Refrigerants containing a colored dye can be successfully analyzed for water using this method.

5.4.2 Water is a harmful contaminant in refrigerants because it causes freeze up, corrosion and promotes unfavorable chemical breakdown. The refrigerants covered in this standard shall have a maximum water content of 10 parts per million (ppm) by weight.

5.5 *Chloride Ions.* The refrigerant shall be tested for chlorides as an indication of the presence of hydrochloric or similar acids.

5.5.1 The test method shall be that described in the Federal Specification for "Fluorocarbon Refrigerants," BB-F-1421B, dated March 5, 1982, (U.S. General Services Administration) section 4.4.4 (silver nitrate reagent). This simple test will detect HC1 and other halogens and requires only a 5 ml; sample. The test will show noticeable turbidity at equivalent HC1 levels of about 25 ppm by weight or higher.

5.5.2 The results of the test shall not exhibit any sign of turbidity. Report the results as "pass" or "fail."

5.6 *Acidity*

5.6.1 The acidity test uses the titration principle to detect any compound that ionizes as an acid. The test requires about a 100 to 120 gram sample and has a lower detection limit of 0.1 ppm by weight.

5.6.2 The test method shall be per Allied approved analytical procedure "Determination of Acidity in Genetron® and Genesolv® Fluorocarbons," GP-GEN-2A (used by permission of Allied-Signal, Inc., Columbia Road and Park Avenue, P.O. Box 1139R, Morristown, New Jersey 07960), or DuPont procedure, "The Determination of Acid Number—Visual Titrimetric Procedure," FPL-3-1974 (used by permission of Freon Products Division E.I. duPont de Nemours and Co., Inc., Brandywine Building 13237, Wilmington, Delaware 19898).

5.6.3 The maximum permissible acidity is 1 ppm by weight.

5.7 *High Boiling Residue*

5.7.1 High boiling residue will be determined by measuring the residue after evaporation of a standard volume of refrigerant at a temperature 50°F [10.0°C], above the boiling point of the sample using a Goetz tube as specified in the Federal Specification for "Fluorocarbon Refrigerants," BB-F-1421B, dated March 5, 1982. Oils and organic acids will be captured by this method.

5.7.2 The value for high boiling residue shall be expressed as a percentage by volume and shall not exceed the maximum percent specified in Table 1.

5.8 *Particulates/Solids*

5.8.1 During the Boiling Range test, a measured amount of sample is evaporated from a Goetz bulb under controlled temperature conditions. The particulates/solids shall be determined by visual examination of the empty Goetz bulb after the sample has evaporated completely. Presence of dirt, rust or other particulate contamination is reported as "fail."

5.8.2 For details of the above test method, refer to the DuPont method for "Determination of Boiling Range, Residue, Particulates" F3200.037.01CW(P) (used by permission of Freon Products Division, E.I. duPoint de Nemours and Co., Inc.).

5.9 *Non-Condensables*

5.9.1 Non-condensable gases consist primarily of air accumulated in the vapor phase of refrigerant-containing tanks. The solubility of air in the refrigerants liquid phase is extremely low and air is not significant as a liquid phase contaminant. The presence of non-condensable gases may reflect poor quality control in transferring refrigerants to storage tanks and cylinders.

5.9.2 The test method shall be that described in the Federal Specification for "Fluorocarbon Refrigerants," BB-F-1421B, dated March 5, 1982, section 4.4.2 (perchloroethylene method). Gas Chromatography, as described in 5.3.3 is an acceptable alternate test method.

5.9.3 The maximum level of non-condensables in the vapor phase of a refrigerant in a container shall not exceed 1.5 percent by volume.

5.10 *Other Refrigerants*

5.10.1 The amount of other refrigerants in the subject refrigerant shall be determined by one of the gas chromatographic methods described in 5.3.3 for the appropriate refrigerant.

5.10.2 The subject refrigerant shall not contain more than 0.5 percent by weight of other refrigerants (see Table 1).

Section 6. Reporting Procedure

6.1 The source (manufacturer, reclaimer or repackager) of the packaged refrigerant should be identified. The fluorocarbon refrigerant shall be identified by its accepted refrigerant number and/or its chemical name. Maximum permissible levels of contaminants are shown in Table 1. Test results shall be tabulated in a like manner.

Section 7. Voluntary Conformance

7.1 *Voluntary Conformance.* Conformance to this standard is voluntary. However, any refrigerant specified as meeting these requirements shall meet all of the requirements given in this standard.

Table 1. Physical Properties of Fluorocarbon Refrigerants and Maximum Contaminant Levels									
	REFRIGERANTS								
	R11	R12	R13	R22	R113	R114	R500	R502	R503
PHYSICAL PROPERTIES Boiling point F @ 29.92 in. Hg	74.9 [23.8]	−21.6 [−29.8]	−114.6 [−81.4]	−41.4 [−40.8]	117.6 [47.6]	38.8 [3.8]	−28.3 [−33.5]	−49.8 [−45.4]	−127.6 [−88.7]
Boiling range °F for 5% to 85% by volume distilled	0.5	0.5	0.9	0.5	0.5	0.5	0.9	0.9	0.9
VAPOR PHASE CONTAMINANTS Air and other non-condensables (in filled container) Max. % by volume	—	1.5	1.5	1.5	—	1.5	1.5	1.5	1.5
LIQUID PHASE CONTAMINANTS Water— ppm by weight	10	10	10	10	10	10	10	10	10
Chloride ion—no turbidity to pass by test	pass	pass	pass	pass	pass	pass	pass	pass	pass
Acidity— Max. ppm by weight	1.0	1.0	1.0	1.0	1.0	1.0	1.0	1.0	1.0
High boiling residues—Max. % by volume	0.01	0.01	0.05	0.01	0.03	0.01	0.05	0.01	0.01
Particulates/Solids—visually clean to pass	pass	pass	pass	pass	pass	pass	pass	pass	pass
Other refrigerants—Max. % by weight	0.5	0.5	0.5	0.5	0.5	0.5	0.5	0.5	0.5

Appendix A
Titles and Sources of Alternate Gas Chromatography Test Methods

ICI
General Chemical Business
ICI Chemicals and Polymer Ltd.
P.O. Box 13
The Heath
Runcorn Cheshire, England WA74QF

Methods for the Analysis of "Arctons," MD1400/32 "Organic Impurities by Gas Chromatography"

Refrigerant	Method Number	Title
R11	RSV/ALAB/CM3 and RSV/ALAB/CM4	Arcton 11
R12	RSV/ALAB/CM5	Arcton 12
R13	RSV/ALAB/CM20	—
R22	RSV/ALAB/CM8	Arcton 22
R113	RSV/ALAB/CM6	Arcton 113
R114	RSV/ALAB/CM21	Arcton 114
R500	RSV/ALAB/CM5	—
R502	RSV/ALAB/CM8	—
R503	RSV/ALAB/CM20	—

(Note: Used with permission of the source.)

DuPont
Freon Products Division
E.I. duPont de Nemours and Co., Inc.
1007 Market Street
Wilmington, Delaware 19898

Refrigerant	Method Number	Title
R11	F3205.165.01CW	Determination of Purity by Gas Chromatography "Freon" 11 Fluorocarbon
R12	F3227.165.01CW(P)	"Freon" 12 Determination of Purity
R13	F3275.165.01CC(P)	Determination of Composition "Freon" 13 Fluorocarbon
R22	F3290.165.01LV(P)	"Freon" 22 Determination of Purity by Gas Chromatography
R113	F3297.165.01CC	"Freon" 113 Determination of Purity by Gas Chromatography
R114	F3305.165.01CC(P)	"Freon" 114 Fluorocarbon—Determination of Composition
R500	F3327.165.01CW(P)	"Freon" 500 Determination of Composition by Gas Chromatography
R502	F3333.165.01CC	"Freon" 502 Determination of Composition by Gas Chromatography
R503	F3337.165.01CW(P)	"Freon" 503 Determination of Composition

(Note: Used with permission of the source.)

Allied
Allied-Signal, Inc.
Engineered Material Sector
P.O. Box 1139R
Morristown, New Jersey 07960

Refrigerant	Method Number	Title
R11	G-11-7A	Determination of Genetron® 11 Fluorocarbon (Assay) Fluorocarbon 12, Carbon Tetrachloride, and Non-Specified Fluorocarbons in Genetron® 11 Fluorcarbon.
R12	G-12-7A	Determination of Genetron® 12 Fluorocarbon (Assay), Fluorocarbons 11, 13, 22 and Non-Specified Fluorocarbons in Genetron® 12 Fluorocarbons.
R13	—	—
R22	G-22-7A	Determination of Genetron® 22 Fluorocarbons (Assay), Fluorocarbons 12, 21, 23 and Non-Specified Fluorcarbons in Genetron® 22 Fluorocarbons.
R113	GSVD-1A	Determination of Genesolv® D (Assay), Fluorocarbons 112, 114, 122, 123 and 1112a In Genesolv® D.
R114	G-114-7A	Determination of Genetron® 114 Fluorocarbon (Assay), Fluorocarbons 113, 115, 123, and Non-Specified Fluorocarbons in Genetron® 114 Fluorocarbon.
R500	G-500-7A	Determination of Fluorocarbon 12, Fluorocarbon 152a and Non-Specified Fluorocarbons in Genetron® 500 Fluorocarbon.
R502	G-502-7A	Determination of Fluorocarbon 22 and Fluorocarbon 115, and Non-Specified Fluorocarbons in Genetron® 502 Fluorocarbon.
R503	G-503-7A	Determination of Fluorocarbon 13, 23, 12, 22 and Non-Specified Fluorocarbons in Genetron® 503 Fluorocarbon.

(Note: Used with permission of the source.)

Bibliography

For additional information on subjects or tests described in this Standard see:

1. ASHRAE Handbook *Refrigeration 1986,* Chapter 7, "Moisture and Other Contaminant Control in Refrigerant Systems." American Society for Heating, Refrigeration and Air Conditioning Engineers, Inc., Atlanta, GA 30329
2. ASTM Standard Designation D3401-78, Standard Test Method for "Water in Halogenated Organic Solvents and Their Admixtures" American Society for Testing Materials, 1916 Race Street, Philadelphia, PA 19103.
3. ASTM Standard D1533-83, "Water in Insulating Liquid (Karl Fischer Reaction Method)." American Society for Testing Materials, 1916 Race Street, Philadelphia, PA 19103.
4. ASTM Standard 2989-74 (reapproved 1981), Standard Test Method for "Acidity—Alkalinity of Halogenated Organic Solvents and Their Admixtures." American Society for Testing Materials, 1916 Race Street, Philadelphia, PA 19103.
5. DuPont Technical Bulletin B-8, "Quality Specifications and Methods of Analysis for the 'Freon' Fluorocarbon Refrigerants." Freon Products Division, E.I. duPont de Nemours and Co., Inc.
6. Parmelee, H. M. "Solubility of Air in Freon-12 and Freon-22" Refrigerating Engineering, June 1951 page 573.
7. Wojtkowski, E. F. "System Contamination and Cleanups," ASHRAE Journal, June, 1964 page 49.

APPENDIX C

1991 STANDARD for

PERFORMANCE OF REFRIGERANT RECOVERY, RECYCLING AND/OR RECLAIM EQUIPMENT

Standard 740

1501 WILSON BOULEVARD • ARLINGTON, VIRGINIA 22209

IMPORTANT

SAFETY RECOMMENDATION

It is strongly recommended that the product be designed, constructed, assembled and installed in accordance with nationally recognized safety requirements appropriate for products covered by this standard.

ARI, as a manufacturers' trade association, uses its best efforts to develop standards, employing state-of-the-art and accepted industry practices. However, ARI does not certify or guarantee safety of any products, components or systems designed, tested, rated, installed or operated in accordance with these standards or that any tests conducted under its standards will be non-hazardous or free from risk.

Price $10.50
Printed in U.S.A.

ARI

FOREWORD

Foreword—This standard applies to equipment for recovery, recycling, and/or reclaiming single refrigerants and their normal contaminants from refrigerant systems. It does not apply to recovery, recycling and/or reclaim from air conditioning or refrigeration systems or storage containers where a mixture of refrigerants exists. In general, technology does not exist to separate refrigerants. No attempt has been made to rate the equipment's ability to remove different refrigerants and other condensable gases from recovered refrigerant. It is the responsibility of the equipment operator to identify those situations where other condensable gases exist and treat accordingly.

TABLE OF CONTENTS

FIGURES

TABLES

PERFORMANCE OF REFRIGERANT RECOVERY, RECYCLING AND/OR RECLAIM EQUIPMENT

Section 1. Purpose

1.1 *Purpose.* The purpose of this standard is to establish methods of testing for rating and evaluating performance of refrigerant recovery, recycle, and/or reclaim equipment (herein referred to as equipment) for contaminant or purity levels, capacity, speed, and purge loss to minimize emission into the atmosphere of designated refrigerants.

1.1.1 This standard is intended for the guidance of the industry, including manufacturers, refrigerant reclaimers, repackagers, distributors, installers, servicemen, contractors and for consumers.

1.2 This standard is subject to review and amendment as the technology advances.

Section 2. Scope

2.1 *Scope.* This standard defines the test apparatus, test mixtures, sampling and analysis techniques that will be used to determine the performance ratings of recovery, recycling, and/or reclaim equipment for various refrigerants. It is not intended to guide the industry in defining required levels of contaminants of recycled/reclaim refrigerants used in various applications. These refrigerants are: R 11; R 12; R 13; R 22; R 113; R 114; R 500; R 502; and R 503 as referenced in the ANSI/ASHRAE Standard "Number Designation of Refrigerants" (American Society of Heating, Refrigerating, and Air Conditioning Engineers, Inc., Standard 34-89).

Section 3. Definitions

3.1 *Recovered fluorocarbon refrigerant.* Refrigerant that has been removed from a system for the purpose of storage, recycling, reclamation or transportation.

3.2 *Recover.* To remove refrigerant in any condition from a system and store it in an external container without necessarily testing or processing it in any way.

3.3 *Recycle.* To reduce contaminants in used refrigerant by oil separation and single or multiple passes through devices which reduce moisture, acidity and particulate matter, such as replaceable core filter-driers. This term usually applies to procedures implemented at the field job site or at a local service shop.

3.4 *Reclaim.* To reprocess refrigerant to new product specifications, by means which may include distillation. Chemical analysis of the refrigerant will be required to determine that appropriate product specifications are met. This term usually implies the use of processes or procedures available only at a reprocessing or manufacturing facility.

3.5 *Standard Contaminated Refrigerant Sample.* A mixture of pure refrigerant and specified quantities of identified contaminants which are representative of field obtained used refrigerant samples and which constitute the mixture to be processed by the equipment under test.

3.6 *Motor Burnout* is the final result of hermetic insulation failure during which high temperatures and arc discharges produce large amounts of carbonaceous sludge, acid, water and other contaminants, and some deterioration of the refrigerant and oil. This can normally be detected by a characteristic burnt smell, and by an acid level in the oil exceeding 0.05 acid number [milligrams KOH per gram refrigerant].

3.7 *"Shall," "Should," "Recommended," or "It is Recommended."* "Shall," "should," "recommended," or "it is recommended" shall be interpreted as follows:

3.7.1 *Shall.* Where "shall" or "shall not" is used for a provision specified, that provision is mandatory if compliance with the standard is claimed.

3.7.2 *Should, Recommended, or It Is Recommended.* "Should," "recommended," or "it is recommended" is used to indicate provisions which are not mandatory but which are desirable as good practice.

Section 4. General Equipment Requirements

4.1 The equipment manufacturer shall provide operating instructions, necessary maintenance procedures, and source information for replacement parts and repair.

4.2 The equipment shall reliably indicate when the filter/drier(s) needs replacement if this method is used.

4.3 The equipment shall either automatically purge non-condensables if the acceptable level is exceeded or alert the operator that the non-condensable level has been exceeded.

4.3.1 The equipment's refrigerant loss due to non-condensable purging shall not exceed 5% by weight of total recovered refrigerant. (See Section 9.4)

4.4 Internal hose assemblies shall not exceed a permeation rate of 12 pounds mass per squre foot [5.8g/cm^2] of internal surface per year at a temperature of 120 F [48.8°C] for any designated refrigerant.

Table 1. Standard Contaminated Refrigerant Sample

	R11	R12	R13	R22	R113	R114	R500	R502	R503
Moisture Content: PPM by Weight of Pure refrigerant	100	80	30	200	100	85	200	200	30
Particulate Content: PPM by Weight of Pure Refrigerant	80	80	80	80	80	80	80	80	80
Characterized by[1]									
Acid Content: PPM by Weight of Pure Refrigerant —mg KCH per kg Refrig.)	500	100	N/A	500	400	200	100	100	N/A
Characterized by [2]									
Mineral Oil Content: % by Weight of Pure Refrigerant	20	5	N/A	5	20	20	5	5	N/A
Viscosity (SUS)	300	150	300	300	300	300	150	150	300
Non Condensable Gases Air Content % Volume	N/A	3	3	3	N/A	3	3	3	3

[1] Particulate content shall consist of inert material and shall comply with particulate requirements in ASHRAE Standard 63.2, "Method of Testing the Filtration Capacity of Refrigerant Liquid Line Filters and Filter Driers."

[2] Acid consists of 60% oleic acid and 40% hydrochloric acid on a total acid number basis.

4.5 The equipment shall be capable of operation to the specifications in ambient temperatures of 50 F to 104 F [10°C to 40°C].

4.5.1 Equipment specified to operate within a controlled temperature range will be evaluated only within that range.

4.6 *Exemptions:*

4.6.1 Equipment intended for a single professional operator and backed by chemical analysis shall be exempt from sections 4.1, 4.2, and 4.3 but not 4.3.1.

4.6.2 Equipment intended for recovery only shall be exempt from sections 4.2 and 4.3.

Section 5. Contaminated Refrigerants

5.1 The standard contaminated refrigerant sample shall have the contents as specified in Table 1.

Section 6. Apparatus

6.1 The apparatus as shown in Figure 1. consists of a 3 cubic ft. [0.085 m^3] mixing chamber with a conical-shaped bottom although a larger mixing chamber is permissible. The outlet at the bottom of the cone and all restrictions and valves for liquid and vapor refrigerant lines in the test apparatus shall be a minimum of 0.375 in. [9.5 mm] inside diameter or equivalent. The mixing chamber would contain various ports for receiving liquid refrigerant, oil, and contaminants as required. A recirculating line connected from the bottom outlet through a recirculating pump and then to a top vapor port would provide for stirring of the mixture. Isolation valves may be required for the pump.

6.2 For liquid refrigerant feed, the liquid valve would be opened. For vapor refrigerant feed, the vapor valve would be opened and refrigerant would pass through an evaporator coil. Flow would be controlled by a thermostatic expansion valve to create 5 F [2.8°C] superheat. The evaporator coil must either be sized large enough to handle the largest system or be sized for each system as required.

6.3 An alternative method for vapor refrigerant feed would be to pass through a boiler and then an automatic pressure regulating valve set at refrigerant saturation pressure at 75 F ± 2 F [23.9°C ± 1.1°C].

Figure 1
Liquid & Vapor Feed Apparatus

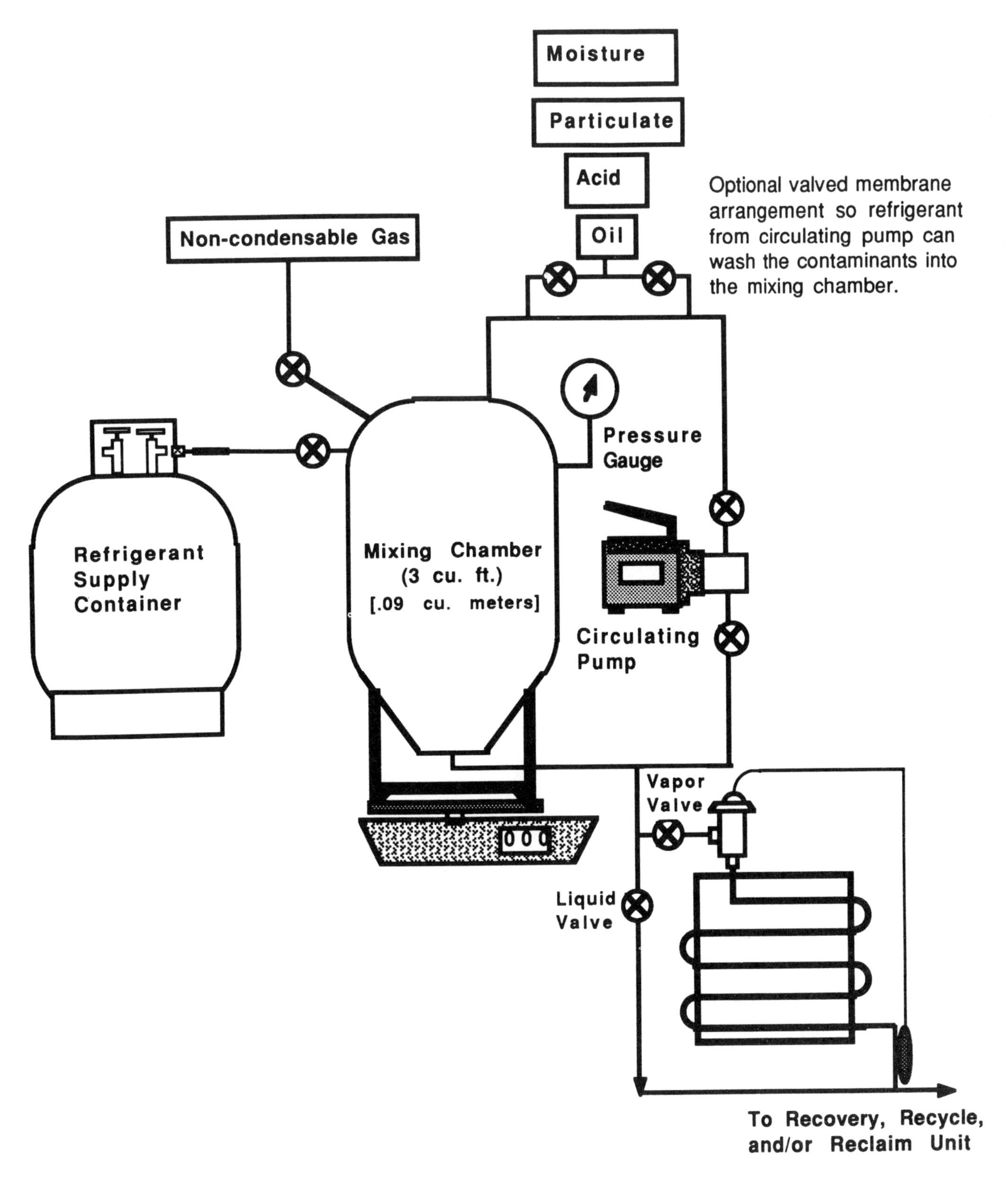

Section 7. Performance Testing

7.1 Contaminants removal and performance testing shall be conducted at 75 F ± 2 F [23.9°C ±1.1°C].

7.1.1 Equipment shall be prepared for operation per the instruction manual.

7.1.2 The contaminated sample batch shall consist of not less than the sum of the amounts required to complete steps 7.1.2.2 and 7.1.2.3 below.

7.1.2.1 A liquid sample will be drawn from the mixing chamber prior to starting the test to assure quality control of the mixing process.

7.1.2.2 Vapor refrigerant feed testing, if elected, shall be processed first. After the equipment reaches stabilized conditions of condensing temperature and/or storage tank pressure, the vapor feed recovery rate shall be measured. One method is to start measuring the vapor flow rate when 85% of refrigerant remains in the mixing chamber and continue for a period of 2 minutes. If liquid feed is not elected, complete Step 7.1.2.3.1.

7.1.2.3 Liquid refrigerant feed testing, if elected, shall be processed. After the equipment reaches stabilized conditions, the liquid feed recovery rate shall be measured. One method would be to wait 1 minute after starting liquid feed and continue for a period of 2 minutes.

7.1.2.3.1 The liquid refrigerant feed test (vapor feed if vapor feed only is selected) shall continue until all the liquid is gone and the equipment shuts down per automatic means or operating instructions. The liquid (or vapor) valve of the apparatus shall be closed and the mixing chamber pressure recorded after 1 minute.

7.1.3 Recycle or reclaim as called for in the equipment operating instructions. Determine processing rate by appropriate means.

7.1.4 Repeat steps 7.1.2.1, 7.1.2.2 (alternately if both elected), and 7.1.3 until equipment indicator(s) show need to change filter(s)

7.1.4.1 For equipment with multiple pass recirculating filter system, change filter(s) and complete recycle or reclaim. Analyze previous batch and current batch after completion of recycle or reclaim.

7.1.4.2 For equipment with single pass filter system, analyze the current batch portion in the storage container.

7.1.5 Refrigerant loss due to the equipment's non-condensable gas purge shall be determined by appropriate means. (See Section 9.4)

7.1.6 Equipment intended for recovery only shall be exept from Section 7.1.4.

Section 8. Sampling and Chemical Analysis Methods

8.1 The recommended referee test methods for the various contaminants are given in the following paragraphs. If alternate test methods are employed, the user must be able to demonstrate that they produce results equivalent to the specified referee method.

8.2 *Refrigerant Sampling*

8.2.1 Special precautions should be taken to assure that representative samples are obtained for analysis. Sampling shall be done by trained laboratory personnel following accepted sampling procedures. The stainless steel test cylinder (approximately 30.5 cu. inches [500 ml] capacity with valves at each end) shall be prepared as follows for obtaining gas and liquid phase samples:

a. Clean test cylinder (with valves) with 0.16–0.67 oz [5–20 ml] portions of reagent grade 1.1.1—trichloroethane or suitable solvents.

b. Blow out test cylinder with dry, <3 ppm water nitrogen.

c. With valves open, place test cylinder and connecting tubing in oven at approximately 230 F [110°C] for one hour.

d. When heated, connect clean copper tubing to storage container and test cylinder. Connect immediately to an evacuation system and evacuate to less than 1 mm mercury [0.133 kPa] (1000 microns).

8.2.2 *Gas Phase Sample.* A gas phase sample shall be obtained for determining the non-condensables. The sample content shall be the minimum required for analysis. Since R 11 and R 113 have normal boiling points at or above room temperature, non-condensable determination is not required for these refrigerants.

8.2.3 *Liquid Phase Sample.* A liquid phase sample is required for all tests listed in this standard, except the test for non-condensables. Do not load the cylinder over 80 percent full at room temperature. This can be accomplished by weighing the empty cylinder and then the cylinder with refrigerant. The cylinder must not become completely full of liquid below 130 F [54.4°C]. When the desired amunt of refrigerant has been collected, close the valve(s) and disconnect the sample cylinder immediately.

8.2.3.1 Check the sample cylinder for leaks and record the gross weight.

8.3 *Water Content*

8.3.1 A liquid refrigerant sample is required. The Karl Fischer Analytical Test Method may be used for determining the water content of refrigerant. This method is described in ASTM Standard for "Water In Gases Using Karl Fischer Reagent" E700-79, reapproved 1984 (American Society for Testing Materials, Philadelphia, PA). This method can be used for refrigerants that are either a liquid or a gas at room temperatgure, including refrigerants R 11 and R 113. An alternate method, Quantitative Determination of Moisture by Karl Fischer Coulometer Titration, may be used. Refrigerants that are a gas at room temperatures with weighed amounts from 1.07 oz to 4.64 oz [30 to 130 grams], of liquid refrigerant shalfl be allowed to vaporize, are to be introduced directly into the anolyte of a Karl Fischer Coulometer.

Refrigerants that are liquid at room temperature with weighed amounts from 1.07 oz to 4.64 oz [30 to 130 grams] of liquid are to be introduced directly into the anolyte of a Karl Fischer Coulometer. For all refrigerants the sample for water analysis shall be taken from the liquid phase of the container to be tested. Proper operation of the analytical and instrumental method requires special equipment, Karl Fischer Reagents, solvents, and an experienced operator. The precision of the results is excellent if proper sampling and handling procedures are followed. Refrigerants containing a colored dye can be successfully analyzed for water using this method.

8.3.2 Water is a harmful contaminant in refrigerants because it causes freeze up, corrosion and promotes unfavorable chemical breakdown. Report the moisture level in parts per million by weight if sample is required.

8.4 *Chloride Ions.* The refrigerant shall be tested for chloride as an indication of the presence of hydrochloride or similar acids.

8.4.1 The test method shall be that described in the Federal Specification for "Fluorocarbon Refrigerants," BB-F-1421B, dated March 5, 1982, (U.S. General Services Administration) section 4.4.4 (silver nitrate reagent). This simple test will detect CL- and other halogens and requires only a 0.16 oz [5 ml] sample. The test will show noticeable turbidity at equivalent halogen levels of about 25 ppm by weight [milligram KOH per Kilogram].

8.4.2 The results of the test shall not exhibit any sign of turbidity. Report the results as "pass" or "fail."

8.5 *Acidity*

8.5.1 The acidity test used the titration principle to detect any compound that ionizes as an acid. The test requires about 0.220 lbs. [100 grams] to 0.265 lbs. [120 grams] sample and has a lower detection limit of 0.1 ppm [milligram KOH per Kilogram] by weight.

8.5.2 The test method shall be per Allied approved analytical procedure "Determination of Acidity in Genetron® and Genesolv® Fluorocarbons," *GP-GEN-2A* (used by permission of Allied-Signal, Inc., Columbia Road and Park Avenue, P.O. Box 1139R, Morristown, New Jersey 07960).

8.5.3 Report the acidity in ppm by weight [milligram KOH per Kilogram].

8.6 *High Boiling Residue*

8.6.1 High boiling residue will be determined by measuring the residue after evaporation of a standard volume of refrigerant at a temperature 50 F [10.0°C], above the boiling point of the sample. A Goetz tube as specified in the Federal Specification for "Fluorocarbon Refrigerants," BB-F-1421B dated March 5, 1982 may be used. Oils and organic acids wil lbe captured by this method.

8.6.2 The value for high boiling residue shall be expressed as a percentage by volume.

8.7 *Particulates/Solids*

8.7.1 A liquid refrigerant sample is required. During the Boiling Range test, a measured amount of sample is evaporated from a Goetz bulb under temperature conditions. The particulates/solids shall be determined by visual examination of the empty Goetz bulb after the sample has evaporated completely. Presence of dirt, rust or other particulate contamination is reported as "fail."

8.7.2 For details of the above test method, refer to the E. I. du Pont de Nemours method for "Determination of boiling Range, Residue, Particulates" F3200.037.01CW(P) (used by permission of Freon Products Division, E.I. du Pont de Nemours and Co., Inc.).

8.8 *Non-Condensables*

8.8.1 A vapor refrigerant sample is required. Non-condensable gases consist primarily of air accumulated in the vapor phase of refrigerant containing tanks. The solubility of air in the refrigerants liquid phase is extremely low and air is not significant as a liquid phase contaminant. The presence of non-condensable gases may reflect poor quality control in transferring refrigerants to storage tanks and cylinders.

8.8.2 Known volumes of refrigerant vapors are to be injected for separation and analysis by means of gas chromatograph. A Parapak Q column at 266 F [130°C] and a hot wire detector are to be used for the analysis.

8.8.2.1 The Federal Specification for "Fluorocarbon Refrigerants," BB-F-1421B, dated March 5, 1982, section 4.4.2 (perchloroethylene method) is an acceptable alternate test method.

8.8.3 Report the level of non-condensables as percent by volume.

Section 9. Performance Calculation and Rating

9.1 The liquid refrigerant recovery rate shall be expressed in lbs. per minute [k/m] and measured by weight change at the mixing chamber (see Figure 1) divided by elapsed time to an accuracy within .02 lbs/min. [0.008 k/m].

9.2 The vapor refrigerant recovery rate shall be expressed in lbs. per minute [k/m] and measured by weight change at the mixing chamber (see Figure 1) divided by elapsed time to an accuracy within .02 lbs/min. [0.008 k/m].

9.3 The recycling rate shall be expressed in lbs. per minute [k/m] of flow and shall be as per ASHRAE 41.7-84 "Procedure for Fluid Measurement of Gases" or ASHRAE 41.8-89 "Standard Method of Flow of Fluids—liquids." If no separate recycling loop is used, the rate shall be the higher of the vapor refrigerant recovery rate or the liquid refrigerant recovery rate.

9.4 Refrigerant loss due to non-condensable purging shall be less than 5%. This rating shall be expressed as passed if less than 5%.

This calculation will be based upon net loss of non-condensables and refrigerant due to the purge divided by the initial net content. The net loss shall be determined by weighing before and after the purge, by collecting the purged gases, or an equivalent method.

9.5 The vapor recovery efficiency shall be expressed in percent and shall be calculated as follows:

E = vapor recovery efficiency (percent)
Psat = refrigerant saturation pressure at 75 F (psia) [23.9°C, kPa]
P = mixing chamber pressure (psia) [kPa] determined in 7.1.2.3.1

$$E = 100.\,(Psat - P)/Psat$$

9.6 The contaminant levels remaining after testing shall be published as follows:

Moisture content, PPM by weight
Chloride ions, PPM by weight
Acidity, PPM by weight
High boiling residue, percentage by volume
Particulates/solid, visual examination
Non-condensables, % by volume

Section 10. Tolerances

10.1 Any machine tested shall produce contaminant levels not higher than the published ratings. The liquid refrigerant recovery rate, vapor refrigerant recovery rate, vapor recovery efficiency and recycle rate shall not be less than the published ratings.

Section 11. Product Labelling

11.1 Type of equipment: (Recovery, Recovery/Recycle, Recovery/Reclaim, Recycle, or Reclaim).

11.2 Designated refrigerants and the following as applicable for each:

11.2.1 Liquid refrigerant recovery rate

11.2.2 Vapor refrigerant recovery rate

11.2.3 Vapor recovery efficiency

11.2.4 Recycle rate

Section 12. Voluntary Conformance

12.1 *Conformance.* While conformance with this standard is voluntary, conformance shall not be claimed or implied for products or equipment within its Purpose (Section 1) and Scope (Section 2) unless such claims meet all of the requirements of the standard.